PRINCIPLES OF YOGA THERAPY

LAURIER-PIERRE DESJARDINS D.O.

Principles of Yoga Therapy

BIOLOGICAL REVOLUTION AND HEALING POWER OF YOGA

Translated by Anita Morf

YogaMasters Press

YogaMasters Press
2278 Montée 2ᵉ Rang
Val-David, QC, Canada
J0T 2N0

Legal Depot
Bibliothèque et Archives nationales du Québec 2014
Library and Archives Canada 2014

Author: Laurier-Pierre Desjardins
World Copyright © YogaMasters Press, 2012

Desjardins, Laurier-Pierre
Principles of Yoga Therapy

ISBN : 978-2-924129-02-9
1. Yoga   2. Alternative medicine

Author : Laurier-Pierre Desjardins

Table of Contents

Note to the Reader

Technical Clarifications

This text is based on philosophical and mystical theory stemming from tantric Yoga. You do not need to master this knowledge to enjoy its benefits. For those who would like to delve deeper into theoretical, scientific or philosophical concepts, I have provided a list of the main works that inspired me, as well as some of the Web sites I consulted, at the end of this book.

Each times a new Sanskrit word or new concept is introduced, it will be in *italics*.

Preamble

This book was written with the aim of inspiring, to guide yogis and yoginîs of the world who may be tempted to embark on a therapeutic adventure.

All practitioners know that Yoga has remarkable curative and transformative powers. This book explores these benefits. It contains superlative assertions about attaining total health at every level of being by diligently applying the suggested principles. These assertions are not only pious hopes, they are based on experience.

If you apply this simple protocol, you will get these results.

Read this book and apply its contents to your life, while helping others do the same. It will enrich your existence, their experience and contribute to greater well-being and quality of life on this planet.

Warning

I am a speaker, a teacher. Not a writer. You will see that in this text. My teachings are usually vibrant, funny, passionate yet full of knowledge. As I was writing this book I found myself often at dismay about my incapacity to convey my passion of Yoga into readable sentences. This is my best shot. I will learn and get better at it.

I believe every yoga teacher or experienced yogis and yoginîs can read and integrate the notions of this book. This isn't the only bold claim I will make - that's my warning. I will often make these claims not because I wrote the book, but because there is so much to learn from the individuals who have shown me physiology in action, and from the Yoga tradition's endless source of wisdom.

Finally dear reader I want to ask you a commitment. I ask you to be bold, intellectually, when reading, go for it! I don't want to speak down to you ; this stuff isn't easy, it exposes the complexity of the human being's organization, and a secular tradition of psychophysical practices that has withstand time to make your self unfold to unbelievable realisations. Do not fear and face it ! This knowledge has tremendous power and could take you to new heights, as it has taken me to new heights.

Foreword

I was delighted, honoured and intimidated when Pierre asked me if I'd write a little "Foreword" to his book. Delighted because I think and feel about Pierre that he is one of my "brothers from another mother". We met a decade ago as participants in the one and only Yoga Research and Education Center's Yoga Teacher Training under the direction of the late Yoga scholar-practitioner, Georg Feuerstein. I was impressed most by Pierre's enthusiasm for Yoga; it's history, philosophies and practices. Indeed, my favourite image of Pierre is him standing at the side of the lecture space, rocking forward and back, seemingly bursting from within with energy and curiosity, asking questions and entering into bustling banter with Georg. Between lectures, I'd often see Pierre offering his healing touch to the other participants (this was one gruelling training!), most famously his "diaphragm release" technique!

I am honoured that he asked me to write this, because Pierre is such a bright light, a real expert in his field, and someone I respect as being an authentic yogi and healer. Honoured, because a man of such experience and knowledge asked *me* to write a foreword to his book. And it is because of these same qualities that I am somewhat intimidated! The man knows the human body like no one else I've met in the yoga community. He certainly knows more about the body than I will ever know! Reading through his text, I was in constant awe once again to witness the depth and breadth of his knowledge.

Whether you agree with Pierre's traditionalist view of Yoga teachings or not, the practices Pierre offers here *work!* We could quibble about *why* they work, but that would be to miss the valuable contribution Pierre makes in offering this erudite, impressive survey of Yoga's deeper practices in a therapeutic context.

So, my delight and enthusiasm override any sense of intimidation I may feel in recommending this book to any teacher, practitioner or even researcher into yogic practice. And I can do so full-heartedly and with gratitude for Pierre's depth of knowledge, skill in teaching, and passionate integrity. This is the book that those of us looking to take the next step in yogic practice and understanding will want to have on our shelves. It will serve us all as a valuable resource.

Poep Sa Frank Jude Boccio
Certified Yoga teacher and Zen Buddhist Dharma Teacher ordained by Korean Zen Master, Samu Sunim. His book, Mindfulness Yoga: The Awakened Union of Breath, Body, and Mind, is the first to apply the Buddha's Mindfulness Meditation teachings to yogâsana practice. Frank blogs at Mindfulness Yoga and Zen Naturalism. Based in Tucson, where he lives with his wife, Monica, their daughter Giovanna and their two cats and two chickens, he travels worldwide, leading workshops and retreats.

Acknowledgements

This book would not have been possible without the inspiration instilled by my revered master, Mahasiddha Kriyâ Babaji Nagaraj. All that is accurate and correct in this book is due to him. Any omissions, mistakes or misinterpretations are my own.

In memory of Maharishi Mahesh Yogi for having sparked my love of Yoga.

In memory of Yogi S.A.A. Ramaiah, my first guru, who initiated me into Kriyâ Yoga.

In memory of Georg Feurstein PhD. who mentored me and brought order into my confused yogic mind.

Thank you to M. Govindan Satchidananda who has guided and mentored me.

My deepest thanks to Frank Jude Boccio, Jessica Robertson, Ted Grand, Anat Van Heiden, Elizabeth McDermott, Swami Swaroopananda, Anne Brissette and Anita Morf for their generosity, patience, kindness, intelligence and competence.

I would also like to thank my wife, constant companion, unfailing source of support and unconditional love, Marie Provost, a true incarnation of Durga on Earth.

Finally, a special thanks to all my students and patients who, over the years, have taught me everything all over again.

General Overview

You may think to yourself: *I am a yogi/yoginî. I am aware of Yoga's immense therapeutic power. I would like to use it to help others. But do I have the knowledge required?*

Let the pages of this book inspire you to explore the answers.

This book is intended for all advanced Yoga practitioners, Yoga teachers, students in training to become Yoga teachers and everyone who would like to broaden their knowledge and explore the therapeutic possibilities offered by this millenias-old tradition.

By the grace of my masters, I was able to distil the essence of numerous techniques that are not easily found in classic yogic literature or modern scientific medical literature.

Over the years, in practicing osteopathic medicine, applying yogic principles and teaching, I have developed recognized expertise in how to free the psychophysical structure and how to best transfer this knowledge to others.

I have seen, treated and taught thousands of patients and students. They have always been and still are my best teachers. I regularly present the knowledge I have acquired in advanced seminars for Yoga teachers and in various osteopathic schools.

I make no claims about self-realization or fully mastering the concepts presented in this book. However, I have put them into practice for over 30 years, using my body and mind as my laboratory for learning and integration. I am not a guru and I am not seeking followers.

For several years now, my students have asked for theoretical support that would incorporate the information I teach in my courses in a practical and accessible document. This book is the result. It gradually leads Yoga practitioners toward a key to self-realization. It serenely and harmoniously takes the physical and mental structure beyond the limits of what our mainstream health and well-being systems consider normal, in fact, raising them to the levels they *should* be. I believe that optimal health and consequently, an optimal psychophysical state, are the privilege and basic right of all human beings.

If you grasp the essence of what I present in these pages and decide to truly apply it to your life, your psychophysical state will be transformed. Regardless of your age, you will experience nothing less than a glorious manifestation of health, energy, power and longevity, in utter joy and happiness! Such is the alchemy of Yoga that is consciously applied.

The first part of the book is *The Functional Anatomy of the Three Bodies*. It links the knowledge of the Yoga Practice with the one of the different anatomical structural levels and their functions.

The second part, *Introduction to Yoga Therapy*, is built on the first. Once the basic knowledge is established, I can move on and explore the therapeutic possibilities of Yoga. So I then come back more deeply on the three bodies. There will sometimes be redundancy between sections of the book; this is done on purpose to ease learning.

May this book be a helpful guide on the path to receiving inspiration, achieving consciousness and developing the innate potential that lies within every human being!

Scientific Arguments

Epistemological Clarification

I strongly believe in taking a scientific approach in studying these phenomena. I will therefore attempt, when possible, to support my claims with scientifically valid sources. Yoga is not always easy to verify scientifically, however. The majority of the techniques explored in Yoga are based on subjective and highly personal experimentation. It is research in the first person, with the researchers themselves as the object of study; as opposed to classical positivist research in which the researchers are supposedly, (although this is now open to debate), detached from their experiment, research in the third person.

In scientific research, when the content is subjective, the most rigorous argumentation and justification methodology remains the one developed for qualitative research. This is a type of scientific study mainly employed in the human sciences (anthropology, ethnology, etc.) and in philosophy (phenomenological method, hermeneutic method), which, among other things, uses triangulation to justify hypotheses. Triangulation helps verify a hypothesis by noting how different authors (usually more than three authors who are well respected in their field) interpret a given subject. Triangulation creates a meta analysis in which the researcher obtains data saturation. This saturation reveals a redundancy of the arguments, in which no relevant new information can be found on the study subject. The subject can then be closed.

Obviously, beyond theories, the best study method in Yoga continues to be practice. Results may vary from one individual to the next. However, these techniques are based on several thousands of years of experimentation, definitely making them worthy of being put to the test!

Vitalism

Finally, I subscribe to the vitalistic school of thought concerning biological phenomena. While remaining strictly scientific, this school believes in a basic principle as a starting point for all biological study: the idea that a vital fluid that is the source of all life animates all living structures. In the yogic tradition, this fluid is known as *Prâna*. It will be the subject of extensive study in this document.

Introduction

Yoga

The Sanskrit word *"yoga"* stems from the Sanskrit root *"yug"*, whose etymological meaning gave *"jugulare"* in Latin and "yoke" in English as well as its derivatives (jugular, to join, junction, etc.). This word therefore means "union", to "unify", "unification". Yoga practitioners strive to reach a state in which the self is absorbed into the divine, a state of complete union with the Light, God, the higher power, where nothing is impossible and time and space are limitless. This ultimate state is attained through *Samâdhi*, or absorption with the subject of the experiment. The practice of *Samâdhi* consists of several degrees and is not easily accessible to everyone. It generally requires an intensity of practice (*tapas*) that is not typical of the classic western yogi.

Modern Yoga was introduced to the Western world toward the end of the XIXth century. Swami Vivekânanda, Paramahamsa Râmakrishna's main disciple, presented this knowledge during the first Parliament of Religions held in Chicago in 1893, during the World's Fair.

Since then, several great masters and their disciples have come to develop their vision of Yoga in the West. These schools are still very prevalent today and continue to develop new, original approaches to Yoga. There is something to suit every taste and personality.

Yoga is extremely popular and accessible, especially in its physical form (Hatha Yoga). If you come across someone who is leaving their weekly "yoga" session, they will often mention the many health benefits the practice brings: well-being, calmness, flexibility, improved physical fitness, etc. However, few practitioners know that Hatha Yoga was originally developed as a physical extension of Tantra Yoga practices.

Tantra

The *Tantras* are a series of philosophical and mystical texts written 1 000 to 1 500 years ago by sages and *rishis* in medieval India. They are the most in-depth yogic texts about humankind's evolution and spiritual achievement. The Tantras are much more recent than the *Vedas*, which date back more than 5 000 years and form a big part of the roots from which Yoga sprang.

The Vedas' four treatises, the *Rigveda*, *Yajurveda*, *Samaveda* and *Atharvaveda*, form a whole. The Tantras could be viewed as their last extension and fifth part. They appear in sacred texts as Yoga's ultimate refinement for humanity's spiritual evolution.

The Tantras are an explanatory and normative model of Reality as experienced by humans. In this sense, they attempt to explain the human experience and suggest ways to optimize it.

The Tantras are not confined, as such, to a religion. They are found as much in the Shaivite texts of Northern India as in the Dravidian *Siddhânta* of the south and Tibetan Buddhism's *Vajrayâna*.

The Tantras are essentially treatises that celebrate the many aspects of divine Energy manifested as static and dynamic matter. This Energy is called *Shakti* in Tantra Yoga, and its dynamic expression in the living structure is *Prâna*. Shakti and Prâna are therefore expressions of Fundamental Intelligence.

In tantric Hinduism, Primordial Intelligence is unchanging, unmanifested, the source of all manifestations. The texts call this source *Shiva*. Non-tantric texts call it *Brahman*.

As soon as Shiva is no longer unchanging, as soon as there is creation, that is Shakti. Shakti is Shiva manifested.

When tantra is incorporated in yogic liberation practices, like adding salt to a dish, it leads yogis and yoginîs to achieve—literally down to the physical structure—ultimate release from the karmic cycle of life and death, also called *Kaivalya*.

This statement implies that tantric practices have a direct effect on matter and the mind. They are very powerful and are always expressed as a manifestation of Shakti (Universal Energy expressed).

In the tantric paths explored, it will come as no surprise that Prâna's fluidity will be altered and the practitioner's quality of life will be modified.

This book therefore explores, from a physiological and therapeutic standpoint, the wonderful transformative possibilities offered by an enlightened Tantra Yoga practice.

These concepts are presented with the aim of healing the psychophysical body. However, as you have probably already realized, it is not necessary to be ill or suffering from any type of dysfunction to begin applying these principles in your life.

Comprehensive Approach

The Yoga tradition forms the ontological framework for this text. The next sections all deal with improving the human structure in accordance with the various levels of being described in classic yogic theory. This theory claims that humans are made up of three to seven sheaths of consciousness, from the crudest (physical body) to the most ethereal (spiritual body). The roots of this knowledge of human nature are found in the Vedas, written over five thousand years ago. But it is in the *Upanishads* and *Vedânta* (literally *the end of the Vedas*) that the concepts were perfected over the centuries, with the refined version appearing in the Tantras.

Three Bodies and Five Sheaths

The yogic tradition, like the majority of philosophical traditions, states that the human being is actually a triune being. It is made up of three fundamental parts or bodies: the gross (coarse) body (*Sthula Sharîra*), subtle body (*Sûkshma Sharîra*) and causal body (*Kârana Sharîra*). Each of these is expressed through more or less crude sheaths or layers of energy. Sthula Sharîra has only one sheath, the crudest, that of the physical structure: *Anna-Maya Kosha* (the sheath made of nutrients). Sûkshma Sharîra has three sheaths: the vital structure, *Prâna-Maya Kosha* (the sheath made of vital fluid), the mental structure, *Mano-Maya Kosha* (the sheath made up of the thinking mind) and the intellectual structure, *Vijnâna-Maya Kosha* (the sheath made up of the discriminating mind). Kârana Sharîra only has one sheath: the spiritual structure, *Ânanda-Maya Kosha* (sheath of bliss).

The theory behind these bodies and sheaths is not originally tantric. It stems from one of the oldest *Upanishads*, the *Taittirîya-Upanishad* (dating back some four thousand years). The Tantras adopted it and it was a perfect fit. I have organized the book based on this yogic theory of the three bodies and five sheaths. Each chapter will focus on one of the three bodies, for a total of three chapters in this first section.

With greater knowledge of the structure and function of each body, it is easy to envision how to treat someone when harmony has been disrupted. This will be the subject of the second part of the book. I suggest ways of planning or applying the therapy based on tools already available to experienced yogis and yoginîs. I am therefore mainly addressing those who have already conducted certain therapeutic experiments on themselves or who would like a specific guide for exploring Yoga therapy.

Note that each body and sheath is closely connected and when a change is made to one of the levels, it will automatically cause an adaptation (transformation) in each of the others. I placed the three bodies in separate chapters to avoid any confusion. But again, all are intrinsically linked and interrelated.

This comprehensive (or holistic) concept conveys a well-known law in *Complementary and Alternative Medicine* (CAM): that the being is a unit and every change made to one part will invariably affect the whole.

This idea of the being's unity is also portrayed by the holographic quality of its expression. That is, each of one body's components contains all of the entire being's components (therefore the contents of all the other sheaths).

This last notion is difficult for the reasoning intellect to grasp. It appears to defy all logic, but is nonetheless a subjective experience felt and shared by the majority of serious yogis and yoginîs.

Yoga Therapy, Âyurveda *and Other Medicines*

In the last years, numerous articles concerning Yoga Therapy seemed to link it with ayurvedic medicine. Ayurvedic practice definitely shares intimate kinship with the historical basis of Yoga. Both have their feet deeply rooted in the Vedas and the oral and written traditions of Hindu philosophy.

I wish however to make a distinction. Yoga is not *Âyurveda*. Yoga, as a philosophy of liberation, is a tradition, which carries a load of psychophysical techniques aiming to unite the different levels of man into a divine whole. This union (Yoga) is ultimately a liberation of his classical shackles, *moksha*.

When these techniques, rituals and *kriyâs* are practised, an undeniable psychophysiological effect of enhancement of the overall shape and state of health is felt. So the corrolary of this practice is a harmonization of physiology. It goes without saying that this yogic methodology could be used in a therapeutic context as a physical and mental medicine. Yoga Therapy thus become, as in all other health practices, a complementary and alternative medicine to the orthodox medical system of our twenty first century society.

Âyurveda is a medicine based on the knowledge of the six classical philosophies of Hindu life (*Shad Darshan*): *Sâmkhya, Nyâya, Vaisheshika, Mîmamsâ, Yoga* and *Vedânta*. So we should not try to blend Yoga Therapy with Âyurveda. As we should not also try to blend Yoga Therapy with *Siddha Vaidyam*, the other ancient medicine of India, at least as old as Âyurveda. Siddha Vaidyam (or the medicine of the siddhas) is the result of the compilation of the alchemical and medical knowledge of South India's Dravidian sages and seers of medieval time. Most of these realized men and women were adepts of Yoga.

Of course each of these approaches could be performed just as another type of Yoga. This is precisely what I am doing with my osteopathic practice and also the same for all my other health professionals' collegues of Yoga.

Yoga Therapy in its most rigorous meaning is the use of classical yogic techniques for therapeutic ends. These techniques are: âsanas, prânâyâmas, dhâranâ, dhyâna, samâdhi, mudras, bandhas and some other kriyâs. This is not an exhaustive list but it constitutes the bulk of the subjects that will be explored in this book.

PART I

FUNCTIONAL ANATOMY OF THE THREE BODIES

CHAPTER 1

STHULA SHARÎRA
THE GROSS (COARSE) BODY

The Physical Sheath
Anna-Maya Kosha

Introduction

In this first chapter, I introduce the physical sheath. It is, of course, the structure that is most easily accessible to Yoga Therapy students.

The Sanskrit etymology of the term Anna-Maya Kosha is eloquent:
– Kosha: envelope, sheath
– Maya: made of, comprised of
– Anna: food, nutrients

Therefore, in Yoga, the physical sheath is *the sheath made of nutrients*. It is not *mâyâ* the illusion, but *maya*, a verb meaning *made of, comprised of*, etc.

This sheath is the crudest or most rudimentary extension of the being's divine essence. In this sense, for the accomplished adept, the physical body is only an instrument, a divine tool. Its essence is immaterial Supreme Intelligence. However, Tantra Yoga teaches that, since the body is an emanation of the divine, it can, in its very matter, reach divine heights and express itself gloriously here on earth. This is what *Sri Aurobindo* envisioned in the last century, as expressed in his and The Mother's Integral Yoga.

The fundamental laws of matter bind the physical structure. That physical layer is comprised of the five basic elements of matter (known in Yoga as *Bhutas*) : Space (Ether), Air, Fire, Water and Earth. It is called the food body because it is made up of and maintained by a continual supply of nutrients from food. This food is also made of these same Bhutas. It is important to specify that diet (which I will cover later) generate the energy needed to maintain the body. This simply means that the physical body has the ability to extract vital energy (Prâna) from food as a means of preservation. It can use its digestive function to accomplish this extraction. Each of the body's functions is simply another tool. Yogis and yoginîs can learn to transcend this function without altering their health...

To understand how this sheath works, I will begin with a short explanation of the fundamental nature of matter. This section is entitled The Nature of Nature.

Then, while listing the reasons that could lead Yoga practitioners to embark on this study, I introduce the essential relationship between structure and function. From there, I unveil the body's physical architecture: the tissues and fundamental ubiquity of the connective tissue.

After exploring the importance of nutrition, especially hydration, in maintaining a healthy structure, I cover the basic concepts and goals of an effective Hatha Yoga practice.

To understand how the physical structure operates, it is necessary to grasp its vast sensory morphology. I will introduce the yogic study of the senses; a study vital to understanding the construct and bodily behaviour of a structure that reacts to the various stimuli to which it is exposed.

This approach will logically lead to concepts concerning the physical sheath's performance: muscle physiology, the essential concept of cocontraction, reciprocal inhibition, flexibility, strength, power and the bandhas.

To complete the basic knowledge required for all yogic practice, it is vital to address the importance of breath during this process. I explore the biomechanical structure of breathing: the diaphragm, its relations and functions. I also introduce the physiology of respiration.

Once this knowledge has been acquired, the physiology of the yogâsanas is established and the transformative potential of the practice, as well as its therapeutic possibilities, begins to emerge.

At the end of the chapter, I present a summary of an appropriate yogâsana practice and reveal some of its secrets.

The Nature of Nature

Organization of Matter and of the Physical Sheath

In the Shaivite philosophy of Tantra Yoga, *Shiva* is *Brahman*, reigning supreme in its omnipotence. Shiva can also be called *God, the Absolute, Light, Supreme Power, Supreme Intelligence*, etc. It is the *Ultimate Principle, the Primordial Essence*. The Tantras combine philosophy and spirituality to explain the genesis of matter. The manifestation of this divine principle (*Purusha*) creates nature (*Prakriti*), from the microcosm to the macrocosm, from the subatomic particle to the entire universe, from the mineral kingdom to the plant and animal kingdoms. *Shakti* expresses this manifestation. Shakti is therefore Shiva's creative energy, Shiva manifested.

Most of the tantric paths are celebrations of universal Shakti. I will touch on this aspect of tantrism throughout the book. Matter is therefore basically an energetic manifestation of ShivaShakti. At the origin of matter there was a creational intentionality. The texts explain this moment as a germinative emission that begins with the primordial sound "*Om*"; the divine vibration, an expression and condensation of the Unchanging Source. This divine manifestation is also the Light Supreme. It is the origin, the genesis of matter and the result of this creativity. *Natarâja*, the dancing Shiva, is an iconographic symbol of this.

The aphorisms below are an attempt to simplify these concepts, which straddle the fence between philosophy and particle physics. Please forgive the epistemological shortcuts.

- Light is the highest basic element and Nature's most subtle. It behaves both in the manner of a particle and a wave. Light is fundamental movement.
- All movement is periodic. It is inspiratory and expiratory, now expanding; now contracting. In this sense, all movement has duality and polarity, pure vibration.
- This vibration produces the primordial sound (*nada*) of Om.
- Condensing vibration creates matter.
- This matter is made up of particles that are more or less solidly held together by polar attraction.
- These particles form the components of atoms. They are called subatomic particles.
- All subatomic particles are themselves made up of *packets of energy, packets of light.* Here we are entering the highly specialized field of particle physics and quantum physics. This exceeds our purpose for the moment.
- One or more protons vibrate in the centre of an atom. Neutrons usually accompany this protonic mass. The proton-neutron whole forms the *atomic nucleus.*
- Electrons gravitate (vibrate, oscillate) around the nucleus, the same number as the protons found in the nucleus. The electronic gravitation around the nucleus forms a sort of dynamic cloud.
- Protons carry a positive electrical charge, electrons carry a negative charge and neutrons carry zero charge.
- The sum total of an atom's charges is therefore zero.
- But when an atom has one electron more or one less than its number of protons, the sum total of the charges is negative or positive.

- When atoms acquire a charge, they are called *ions*.
- Negative ions are *anions* and positive ions are *cations*.
- Ions join at their opposite polarity.
- As of 2012, there are 118 different known atoms in Nature, making up the basic elements of creation. These are grouped and analyzed in the *Periodic Table of Elements*.
- When two or more ions or atoms join, they form a *molecule*.
- Molecules also have total charges that can be negative, positive or neutral. However, the majority of the physical body's molecules are polarized.
- A molecule can be very small, like a water molecule (H_2O), in which 2 atoms of hydrogen bind with one atom of oxygen.
- A molecule can also be made of several hundreds, if not thousands of atoms. These are called macromolecules. Hemoglobin (Hb) is one example. It is made up of 6 268 atoms distributed as follows: $(C_{738}H_{1166}FeN_{203}O_{208}S_2)_4$.
- Nature is made up of basic elements that are essentially in molecular structure.
- Molecules therefore form the fabric, the building blocks of all matter, whether organic or inorganic.
- The mineral kingdom is made up of more or less complex molecules in solid, liquid or gaseous form.
- The plant and animal kingdoms are also made up of molecules. However, they form structures (*cells*) that are a bit more complex and are capable of supporting *life*.
- Cells are the smallest *unit of life* in nature. It is at this level of matter that Shakti becomes Prâna, animating the cell with life.
- It is therefore at this primordial stage that all metabolic activities that characterize living organisms, from the simplest to the most complex, take place.
- Cells are made up of molecules that form their peripheral membrane and inner content.

– Each cell contains an *intracellular fluid* called cytoplasm (a type of gel consisting predominantly of water), which bathes other molecules called *organelles* (small organs). These operate as small systems that contribute to cellular metabolism.

– The cytoplasm and organelles form the *cytosol*.

– The *cellular nucleus* is located in the centre of the cell and contains various forms of genetic material belonging to the host. This material forms the *nucleic acid* molecule (*DNA*).

– The nucleus also contains another centre, the *nucleolus*, which holds a copy of the DNA, the nucleic acid *RNA* used for metabolism and cellular reproduction.

– The main molecule families that play a role in human cellular metabolism are: *carbohydrates, proteins, lipids*, nucleic acids and water.

– All molecules can become more complex by linking to other molecules according to need or cellular function.

– Determination of the biological individual's needs appears to be governed by the expression of part of the genetic code programmed in the DNA molecule. This forms the basis of genetic science.

– Epigenetics is a recent biological science that sheds light on a new aspect of ontogenic determinism: that genetic expression would be linked or subject to the nature of the cell's immediate environment.

– So the expression: "genes engender development" then becomes: "environment modifies genes that engender development" with environment playing a key role in biological evolution.

– Molecules can also bind with any other element (atoms, ions or molecules). This is why every element in the periodic table (and therefore the universe) can be found in every individual, in traces or significant amounts.

– When two or more similar cells join, they form a tissue.

– Tissues combine to form organs.

– Organs join to create systems.

– All systems form the plant or animal organism.

Human beings are probably the most accomplished creation in the animal kingdom. The major argument supporting this hypothesis is that humans are the only creatures capable of transcending their physiological processes. We will discuss how Yoga enables us to do so.

Yoga, Health and Knowledge

Yoga and Total Health

Yoga has the ability to generate perfect health. Not only the health of the physical body, but at every level: physical, vital, mental, intellectual and spiritual. Tantra rishis and sages have developed numerous yogic techniques (*kriyâs*) to achieve this.

It is under the *Hatha Yoga* corpus that these kriyâs were first developed for the physical structure, in order to complement good nutrition. We will come back to this later.

Three Ages of Practice

Before continuing, let us look at the individuals who practice Yoga. We will then see how health can result. When the world is not in too much turmoil, there are three ages of Yoga practice:

1) Children up to adolescence, usually around the first 12 years of life;

  – At this age, the child's Vital Energy appears gratuitous.

  – Yoga is automatic, as is health.

  – The child is in the *power of now.*

  – Children appear to lose these faculties and gradually forget during the onset of adolescence.

2) Add 30 to 50 years;

– A time during which the adult learns or relearns to control the flow of vital energy through Yoga or a healthy lifestyle.
– Health is then manifested or returns.
3) Subsequently;
  – Maintenance or dissolution of the physical structure according to one's destiny.

The preceding is often the subject of tiresome philosophical and sociological debate. My intention goes beyond futile argument. I subscribe to action Yoga, Kriyâ Yoga, and believe that regardless of age, health and self-realization can be achieved through Yoga.

Knowledge of the Physical Sheath

Anyone who is serious about practicing a physical therapy first needs to gain sound knowledge of the human body, the physical structure. In every field of medicine throughout history, students began their training with a study of the body's anatomy. If you are a student of Yoga therapy, I recommend that you do the same.

Why is it so important? It is vital that all therapy practitioners build a mutual relationship of trust with their patients, a helping relationship. The help that is given will stem from the professional expertise acquired by the therapist, who must demonstrate ability and competency. These qualities not only require psycho-emotional maturity to confidently enter a helping relationship, but also a certain mastery of the Yoga techniques based on solid anatomo-physiological knowledge. Therapists are supposed to *know*. It is the main reason they are consulted. Their science and wisdom are the two things that draw people to the therapy.

No one can give you wisdom. It comes with maturity and clinical experience. However, you can acquire greater knowledge to add to your Yoga science. This book will help.

With patience, your practice will become easier and more comfortable. You will gain confidence. You will gradually develop therapeutic intuition, which will make you even more powerful and effective. Therapists (healers, shamans) possess *Knowledge (Vidyâ)*. They know what to do or how to proceed to bring about the desired transformations.

This may seem like a gargantuan task and in some regards, it can be. But you can start the process without getting bogged down by the details.

What counts, in the beginning, is an understanding of the physical body's architecture, its makeup and structure. The rest, as you will see, will come by itself, with your passion in discovering this infinite universe.

Living Anatomy
or Brief Reminder of Human Biology

Yoga Therapy and Functional Anatomy

Our study will be a little more original than classic anatomy. Without rejecting the latter, what is useful for Yoga therapists is living anatomy.

The classic study of anatomy in medicine is based on the dissection of cadavers. Most of our anatomy textbooks and the pictures they contain therefore use an inanimate being to explain physical structure. Of course, in medicine, this information is completed by physiology, the science of how the anatomical structure works.

I will therefore accentuate our anatomical study by making it directly functional. That is, I will essentially produce a study of the living human structure, a *functional anatomy*. For those who would like to further explore the nomenclature and geographic locations of the anatomical structures, there are countless classic anatomy books that are amply illustrated and perfectly adequate. There is no need to reinvent the wheel on this subject.

Functional anatomy's main characteristic is that it makes way for *life* in its expression. And the greatest expression of life in the physical sheath's structures is *movement.*

In a living body, everything is in motion. This mobility, this intrinsic motility of the structures, characterizes functional anatomy. It is therefore an anatomy in movement, just as Hatha Yoga is comprised of a series of asânas performed through movement.

The Physical Sheath as a Structure

The body's physical structure is what holds the crude shape of your being. The human body and its components are made up of material structures that form a shape.

According to biological science, determination of bodily form appears to be governed by genetic and environmental impulses. Genetic determinations are found in the cell nucleus at the heart of the initial cell, the *zygote* (first cell of the embryo and therefore the first cell of the individual to be born) and each subsequent cell in the developing organism. However, this aspect of biological knowledge is still vague in many respects and determinations of form (morphogenesis) remains essentially speculative even though several paths have been actively explored in research. The concept of *form* and *morphogenesis* is found in yogic philosophy, conveyed by the notion of *Rupa*, a Sanskrit term that encompasses this study. There is therefore a yogic exploration (via various concentration and meditation exercises that I will cover later) of the form of things and Nature, providing access to a more fundamental knowledge of matter and the being.

Structure and Function

Biologically, in the sphere of the living, cells and tissues express this body structure. That is all. It is both simple and complex.

To be functional in life, this structure must always remain stable and fluid. Remember that the human structure is living, animated. This leads to the expression of our first axiom (*sutrâ*): *Structure Governs Function.*

This statement is one of the basic principles underlying traditional osteopathic medicine.

Like any axiom, this sûtra or koan must be meditated on to reveal all the wisdom it contains. No matter the structure: material or immaterial, physical or conceptual, organizational or pertaining to a building, it has a function. If the structure is stable, the function can express itself more easily. If it is too rigid or unstable, function will be affected. The opposite is also verifiable: *function determines structure…*

The same is true for inanimate matter and living matter. Structure is the same for all matter, a combination of molecules forming various architectures. The physical body therefore has a structure with the characteristic of being animated with life, with movement.

Architecture of the Physical Sheath: The Cells

The physical body's matter is made up of molecules grouped in more or less globular clusters called *cells.* In Nature, cells are the smallest units to hold life. Their components, intracellular organelles, make up as many small organs as are needed to carry out the roles for which they are programmed.

The inside of each cell is highly fluidic. This intracellular fluid, also called cytoplasm, bathes all the organelles. The organelles and cytoplasm make up the *cytosol* of each cell.

One of the main roles of all cells, their primary goal in life, their *metabolism*, is to produce various proteins in their cytosol. In biology, this is called *protein synthesis.*

The hundreds of billions of cells in the human body each have various metabolic functions, but protein synthesis is one of their major tasks.

Proteins are complex macromolecules that form the structure's basic building blocks. They are also the basic components of the enzymes, hormones, transmitters and various metabolic reaction modulators, as well as most of the substances flowing in the living body's fluids (like blood and lymph). They must constantly be renewed.

Protein synthesis is determined by genetics and environment for each type of cell. This vital cellular function enables the constant repair and regeneration of the cellular structure itself, by replacing worn elements as well as producing all other components needed to keep the organism running smoothly.

Architecture of the Physical Sheath: The Tissues

A group of similar cells is called a tissue. Classical anatomy identifies four basic tissues that make up an animal organism: *epithelial* tissue, *muscle* tissue, *nervous* tissue and *connective* tissue. These tissues are differentiated very early on in the embryo's development. It is said that after three weeks of gestation, all of the growing organism's tissues are already differentiated and established. Each tissue forms a set of similar cells that carry out the same functions and work toward the same metabolic goal.

Each organ in our physical being is a combination of these four tissues. The main anatomical difference between the organs in the human body is in their concentration. Now, let us examine some of the characteristics of these tissues.

Epithelial Tissue

Epithelial cells are configured in a typical manner. This tissue's cells are the most closely packed and tightly woven. There is very little space between the cells (called *interstitial space*). The epithelial tissue, or *epithelium*, is so compact that water has difficulty passing through, making it somewhat watertight. The cells are pressed together, often in several layers.

Epithelial tissue is found on the surface of the skin (*epidermis*) and as the main lining, called *mucosa*, in the body's cavities.

Epithelial cells (or *epitheliums*) are also responsible for secretions on the surface of this mucosa, making the surface of the epitheliums moist with their fluids. This epithelial tissue therefore has specialized cells that produce and secrete all of the organism's mucus and digestive juices.

The epithelial tissue also forms the functional tissue of the body's dense organs, or the *parenchyma* of these organs. The liver's parenchyma, for example, is made up of the liver's functional cells (*hepatocytes*). One could also talk about the parenchyma of the kidneys, lungs, etc.

Muscle Tissue

Long and tubular, or fusiform, the cells of the muscle tissue are called *muscle fibres* or *myocytes*. There is more interstitial space between the myocytes than in the epithelial tissue. The myocytes have a very original quality, the ability to *contract*, due to a very orderly configuration of contractile proteins in their cytosol. These proteins are so abundant within the myocytes that they fill the entire volume of the cells, pushing all of the other classic organelles to the edge, crushed at the sides of the cell membrane.

When triggered by a nerve impulse, essentially an electrical current, the structures of these proteins (*actin, myosin, troponin, tropomyosin*, etc.) are modified and slide between one another. The net result is that the cell itself shortens and becomes denser. Through summation in the muscle tissue, the latter contracts and increases in density. Muscles in a state of contraction will often be shorter than when at rest, but will always be denser, more tonic. In this state, the muscle has a certain *tone* that is greater than when at rest.

There are three types of muscles in the human body:

Skeletal muscle:
— The most abundant
— Connected to the bones and therefore responsible for movement through space
— Voluntary contraction by nerve stimulation

Smooth muscle:
— In the walls of hollow organs and around secreting cells
— Responsible for visceral motility and propulsion of the contents
— Involuntary contraction by nerve and hormone stimulation

Cardiac muscle:
— Forms the myocardium of the heart
— Responsible for the pumping action of the heart
— Involuntary contraction by nerve and hormone stimulation

I will come back to muscle physiology later.

Nervous Tissue

This tissue is a bit more complex than the two previous ones. It forms the basis of the *nervous system*. The interstitial space is quite large.

The most well known nerve cells are the *neurons*. Neuronal morphology is important because it determines its specific function. Each neuron has a cellular body, a *soma,* whose walls are often lengthened by hundreds of small excrescences radiating around. These excrescences, called *dendrites,* form the neuron's receptor area. The neuronal soma also has one or two very long excrescences, called *axons.* Axons make up the neuron's transmission area.

Neurons have *conduction* abilities. They can produce electrical currents or *nerve impulses.* These impulses are transmitted by the neuron's axons toward the dendrites of neighbouring axons and from one neuron to the next in the central nervous system. This transmission forms a continuous chain in which information can be sent to the various areas within the organism.

There are approximately one hundred billion neurons. Their somas form what is commonly called the *grey matter* of the central nervous system.

Neuron metabolism is assisted by another group of cells whose main role is to support neuronal activity. These cells, called *neuroglia* or *glial cells*, play a protective, nutritional and supportive role. They are located all around each neuronal soma in the central nervous system. There are often two, three or even up to ten or more neuroglia for a single neuron. In their nutritional role, they filter nutrients from the many capillaries so that only *glucose* is allowed to reach the neurons. This simple sugar, or *carbohydrate* is the neuron's only food and main source of energy. The neuroglia's filtering, nutritional and protective function constitutes what is commonly called the *blood-brain barrier*.

The central nervous system is made up of the *brain* and *cerebellum* inside the skull as well as the *spinal cord* inside the vertebral canal of the vertebral column.

The axons that extend the neurons are the main pathways for nerve impulses. All of these pathways form the nervous system's *white matter*.

Unlike the neuronal soma (or grey matter), which is located in the central nervous system, some axons originate from the centre and extend to the periphery, toward the rest of the trunk or limbs. They emerge from the spinal cord and spinal canal through the intervertebral foramen at each level of the spinal column, between two adjacent vertebrae. This group of axons is called a *peripheral nerve*, a nerve to the right of the column and another to the left. There are eight pairs of cervical nerves, twelve pairs of thoracic nerves, five pairs of lumbar nerves, five pairs of sacral nerves (passing through the sacrum) and one pair of coccygeal nerves (emerging from the coccyx). There are twelve pairs of cranial nerves that emerge from the skull, mainly from the base of the brain (*brainstem*). They essentially serve the various organs of the head. The tenth cranial nerve (*Vagus nerve*) is the only one that is distributed all the way down into the abdominal cavity.

The nerves make up the *peripheral nervous system.*

Nerve impulses are electrical currents produced by the neurons. They travel from one neuron to the next. When an impulse runs in an axon and travels in a peripheral nerve, its goal is to reach a muscle. This results in the muscle's contraction, making it a *motor impulse.* The centrifugal direction of this impulse, from the centre (central nervous system) to the peripheric muscle, is said to be *efferent*, or an *efferent impulse.*

However, when the impulse originates at the periphery and travels along a nerve toward the central nervous system, it is moving in the opposite or centripetal direction. This is called *afferent*, or an *afferent impulse.* Afferent impulses are not motor impulses, they stem from the periphery, usually from a sense organ. Each organ associated with one of the the five classic senses contain sensory nerve fibres that, when stimulated, trigger afferent impulses and send sensory information along one or more nerves toward the central nervous system. There, the information can be integrated and analyzed in order to trigger an appropriate return response.

The peripheral nerves can therefore carry efferent motor information or afferent sensory information. Some nerves are mixed and carry both types of impulses. Others are solely sensory or solely motor nerves.

The motor part of the peripheral nervous system sends impulses to the muscle structure. If the target is a skeletal muscle, its contraction will be voluntary. The motor part of the peripheral nervous system that is in charge of the voluntary contraction of the skeletal muscles is called the *somatic nervous system.*

However, if the impulse is directed at a smooth muscle, its contraction will be involuntary. The motor part of the peripheral nervous system that is in charge of involuntary muscle contraction (smooth or cardiac muscle) is called the *autonomic nervous system*. This system has two components that are usually complementary: the *sympathetic nervous system*, which has thoracolumbar outflow and is capable of controlling warning and emergency situations, and the *parasympathethic nervous system*, which essentially has craniosacral outflow and is mainly in charge of conserving the organism's energy and maintaining its resting state.

Since the autonomic nervous system is a motor system that acts involuntarily, it generally acts at a *subconscious* level. Through various Yoga kriyâs, yogis will be able to gain some control over this system.

Cerebral Functional Anatomy

The brain is an incredibly complex organ, whose different parts are continually under intense study. The majority of the cerebral regions were identified and mapped during the XXth century, in large part due to the work of Dr. Penfield in Montreal. Each was found to have a specific regulatory or integrative function or a functional correlation with a precise area of the body. A certain lateralization of the cerebral regions was also brought to light. Therefore, a region in the right hemisphere will not operate exactly like its matching region in the left hemisphere.

In another epistemological effort to simplify the data, the brain can be roughly divided into four large functional regions of nervous tissue:

- **The left brain**: located in the *cortex* (the brain's peripheral layer) of the left hemisphere. It is mainly in charge of analysis, logic, detail, calculation and reason.
- **The right brain**: located in the cortex of the right hemisphere. It is mainly in charge of symbolism, imagery, creativity, the big picture, intuition.

> – **The limbic brain**: located under the previous two, at the centre of the brain. It is in charge of emotions and drives.
> – **The brainstem**: located between the brain and spinal cord, at the *foramen magnum* of the occipital bone. It is in charge of organizing all autonomic motor activities. It is an important centre for the control and distribution of information to the other regions.

The left brain essentially works at a conscious level. Its reflection and analysis provides the understanding needed to motivate action. The three other broad regions are also involved in carrying out action and behaviour, but are mostly subconscious. Their work is therefore carried out unconsciously. When these four parts flow smoothly, the action is harmonious, powerful and efficient. Unfortunately, harmony between these four regions is easily disrupted, which sometimes leads to self-sabotage and failed undertakings.

A visualization can help explain how the brain perceives outer reality.

Imagine a forest spread out before a cerebral observer. The left brain interprets what it sees as trees, branches, leaves, shrubs, flowers and animals. It identifies these things, classifies them and takes in all the data about the forest.

The right brain rises above this and sees the forest as a whole. In one glance, it perceives its colour, depth, hues, coolness, odours, sounds, the wind in the leaves and the creaking of tree limbs. It drinks in the forest's energy and vibration.

The limbic brain perceives the experience as being either pleasant or unpleasant. The forest's darkness and depth frightens this part of the brain. It feels crushed by the forest's density, suffocating humidity, fog and lack of light. Inversely, the forest can trigger a feeling of joy; due to its open space, the beautiful sunbeams piercing through to the forest floor, the small curls of ethereal mist draped over the majestic leaves dripping with dew, etc.

However, it is the brainstem that will signal the brain to actively begin exploring the trails and allow the physical structure to coordinate its sensorimotor functions to be able to seize, touch, take, walk, run and cross the forest.

In the second chapter, we will look at Yoga tools that provide remarkable control of the brain and subconscious.

The Key to Physical Structure

Connective Tissue

The connective tissue is the least known of the tissues that make up the physical sheath. However, in a study of functional and yogic anatomy, it is probably the most important. It is through this tissue that our physical structure and all of its components can be reached.

In fact, it is the connective tissue that links all the tissues that form singular groups of organs and systems. It is the most abundant tissue in the body. It *connects* everything to everything, from cells to other tissues, organs and systems. All of the tissues and organs, all of the spaces and cavities are surrounded and filled with connective tissue.

Collagen Fibres and Ground Substance

Like the other tissues, the connective tissue is made up of a set of more or less loosely grouped families of similar cells. It is the tissue whose interstitial space is the vastest. In fact, there is so much space in certain connective tissues that their cells have no direct contact. The interstitial space is obviously not empty. It is usually filled with protein fibres and fibrils that are bathed in a more or less aqueous gel (over 80% water). The fibres of the connective tissue essentially consist of the protein filament *collagen*. The fluid and fibers as a whole (the *matrix* or *ground substance*) create a link between the cells scattered in the connective tissue. By its nature, this tissue is very humid, sometimes even liquid and flows dynamically inside the human organism. The connective tissue's more or less fibrous and aqueous ground substance bathes the cells of the tissue in the interstitial space. It therefore forms a sort of liquid or, more precisely, a fluid, around the cells. This is called *extracellular fluid* as opposed to *intracellular fluid*, which bathes the cell's organelles.

An important mechanical property of the connective tissue is that its extracellular fluid or matrix, by virtue of being chiefly aqueous, is by and large incompressible, a little like water. Moreover, its collagen fibres are very resistant. They are twisted protein, more or less agglutinated by type of connective tissue, not very extensible and thus highly resistant to all forms of traction while retaining a bit of flexibility (a little like ropes). Moreover, they are insoluble in their gel. These properties give all the connective tissues a remarkable biomechanical plasticity.

Where do these collagen fibres that form the basis of every physical structure come from? They are produced by the very metabolism (protein synthesis) of the connective tissue's cells. These are the connective tissue's most abundant cells, called *fibroblasts*. Fibroblasts assemble and gather the proteins into microfibrils and then expel them outside the cell (*exocytosis*) into the tissue's extracellular fluid. Depending on the intensity of the fibroblastic metabolism, the tissue will be more or less fibrous. The connective tissue's cells are therefore caught in the mesh of fibrous tissue they themselves have created. A little like a spider in the middle of a web it has woven.

Types of Connective Tissue

Just like there are various types of muscle and epithelial tissues, there are several types of connective tissues. The types mainly correspond to the different concentrations of fibres and ground substance in the tissue. They include:

- **Loose connective tissue**, which fills all the cavities between the organs and wraps them in a sort of fibrous web. The *fascia* belongs to this category.
- **Dense connective tissue**, whose fibres run in a more regular direction, giving this *structure* more rigidity. *Ligaments, tendons* and *articular capsules* belong to this group.
- **Bone tissue**, in which the mineralization of the collagen fibres gives the skeleton its structure. Each bone is made up of various *osteocytes (osteoblasts and osteoclasts)*, surrounded by mineralized collagen fibres.
- **Adipose tissue** whose cells *(adipocytes)* are filled with giant drops of lipids and form the basics of the organism's fats and energy reserves.
- **Cartilage tissue**, whose main cells are *chondrocytes*. This is a dense fibrous tissue whose fibres and cells are buried in a gelatinous matrix that is smooth and elastic, called *chondroitin sulfate*.
- **Vascular tissue**, *blood* and all other circulating fluid are considered connective tissue. They are all made up of cells bathed in a more or less fibrous fluid matrix. In the blood, these elements and cells are *erythrocytes, leucocytes, platelets* etc. The fibers are mostly *fibrin* and the fluid matrix is *plasma*. This is obviously the most fluid and flowing connective tissue. The *lymph* is another. *Cerebrospinal fluid* also falls into this category.

Integrins and the Cytoskeleton

Connective tissue spreads to all levels, as far and deep as the cellular level itself. Each cell (not only of the connective tissue, but all tissue) is in complete continuity and physical communication with its immediate environment via small, collagen-type protein fibrils. These proteins, called *integrins*, pierce the cell membrane of each cell and link the extracellular fluid with the intracellular fluid. Integrins are found everywhere, on and through the cell membrane of every cell and throughout the intracellular cytosol, even piecing the nucleus. It is the same for all tissues, all organs and all systems.

These integrins act as cell receivers and emitters, channels for transmembrane exchanges, structure proteins, etc.

Inside each cell, the cytosol contains protein filaments that spread through the cytoplasm in every direction and come into contact with the transmembranal integrins. These filaments are also called *tubules, microtubules* and *microfilaments* in the cell. The whole makes up the *cytoskeleton*, the skeleton of the cell.

In short, the cytoskeleton is made up of filamentous collagen-like proteins. These proteins, this cytoskeleton, form the framework, the internal latticework of all the cells of every tissue. The cytoskeleton is also in direct contact with the extracellular fluid of every tissue via the transmembrane integrins.

The Secrets

Let us now look at some of the connective tissue's lesser-known characteristics:

- Collagen and all structural proteins in the connective tissue stretch and loosen in the presence of *heat.*
- This connective tissue exhibits a behaviour called *tensegrity* that is well known in the field of architecture (see below).
- The connective tissue's cells and collagen fibres are set up in a very orderly fashion. At first glance, its components will often appear random and chaotic. But this is a two-dimensional illusion. There is order in the chaos. The connective tissue's three-dimensional structure shows that the components are actually highly ordered. Through the microscope, the design closely resembles a *crystalline* molecular structure (see below).

Heat

All movement is fuelled by energy previously stored in the organism. When energy is used, heat is produced. This is characteristic of molecular exchanges that occur during movement. If this mobility also takes place in a hot exterior environment, the movement will be more fluid and smooth because the mobilized tissues will slide more easily over one another. Hatha Yoga was developed on the Indian subcontinent, where it is hot year-round. It is therefore not surprising that yoga sessions are even more fluid when practiced in a hot exterior environment. Bikram, Moksha Yoga or "Hot Yoga" classes are now very popular in the Western world. This type of environment further supports Yoga's glorious transformation of the physical structure.

Tensegrity

Tensegrity is an architectural structure's inherent ability to adapt, morphologically, to internal or external forces and stresses. This helps save energy and protects structural integrity. The physical connective structure's highly adaptive biomechanical capacity is a marvel of tensegrity.

Crystal

As for the crystalline design of the connective tissue's fibres and cells, let us cross into the field of physics.

The molecules of solid or liquid crystals have similar known properties. Each linking electron on the molecule's outer energy layer (*valence*) is mobile and shared by all the other molecules in the system (in this case, the crystal). Therefore, the electron circles the crystal. The movement of the electrons around the system is thus an energy continuum in motion in the crystalline structure.

Piezoelectricity

Let us continue our physics lesson.

Due to the movement of the valence electrons around this liquid crystal that makes up our fibrous structure, our connective tissue, the crystal molecules become semi-conductors. One of the most important physical properties of crystalline semi-conductors is *piezoelectricity.* It produces an electrical polarization, an electrical field, when the crystal is exposed to mechanical stress.

Electrical fields appear around all living tissue exposed to compression stress, stretching, twisting, etc.

Therefore, forces acting on the tissues stimulate electrical currents and the apparition of energy fields on the ionic molecules of the physical structure's humid (or liquid) crystal. These bioelectromagnetic fields could not exist without their essentially aqueous environment (see the Importance of Hydration, a little further).

Living Matrix

The connective tissue is therefore a living continuum reaching every cell and tissue. It forms a whole that connects every part of the organism. The result is a *living matrix.*

This more or less fibrous and fluid matrix is the human physical structure's ultimate information and communication system, of which the nervous system and the senses only form a part. Hatha Yoga makes this system perfectly efficient.

Final Secret

Biologists have long believed that the connective tissue was only a relatively non interesting set of inactive fibrous wrapping.

However, toward the end of the XXth century, researchers studying the fascias made remarkable advances in understanding the connective tissue's properties.

They discovered that the tissue's main cell, the fibroblast (which produces collagen fibrils) has the amazing ability to become contractile.

Fibroblasts are found in variable concentrations in all of the connective tissues. Their cytoplasm holds some proteins from the actin family, which are mainly found in the myocytes (muscle cells). These proteins are contractile. That is, when subject to certain stimuli, they can change their morphology and shrink.

When exposed to certain types of stress, fibroblasts can contract, a little like smooth muscle cells. This could, among other things, explain the connective tissue's frequently observed tendency to wither or contract (for example, its active role in scars). The more exact name for these cells is *myofibroblasts*.

Nutrition

The Importance of a Healthy Diet

Anyone who consciously decides to adopt a healthy lifestyle will inevitably encounter information about the foods he or she should be eating. Our current Western world is a consumer society, in which food products are viewed as a means of generating profits rather than supporting the health of the consumer.

We have always known that diet plays a major role in health. In fact, Hippocrates said *"Let food be thy medicine"* over 2500 years ago. Today, there are countless diet theories to choose from, all promising health and energy with diligent application. Some are based on spirituality, others on life philosophy, still others on nutritional science. Several contradict each other. However, the contradictions are often due to differing viewpoints, separate paradigms. There is no sense comparing apples and bananas.

Most authors of Yoga-related books favour a vegetarian diet. Vegetarian food is quite varied and can be adapted to almost every society. And yet, Tibetans (including His Holiness, the Dalai Lama) are not vegetarian. Different paradigms, different solutions. The danger in nutrition, like all other human enterprise, is blind or forced adhesion to a belief that subordinates reason and logic and imposes sectarian behavior robbed of all meaning.

If you are able to extract Prâna from the foods you eat and absorb it in your physical structure, your diet will nourish you and be a true source of energy, vitality and health. This process calls not only on your digestive capacity, but also your culture, which plays a significant role in your dietary beliefs. In Chapter 2, I will explain the need to clear the mind in order for the physical body to operate at an optimal level.

All that being said, it is still possible to establish a few characteristics of healthy nutrition. There is one constant: it takes effort. To properly nourish the physical structure, you need to *do* so. You have to rise above what is easy and overcome laziness or procrastination.

Characteristics and Advices

— **Eat sparingly**: You may have already heard something along the lines of *"Your main meal should fit in one hand"*. Before applying this, you would need to determine if it would meet all of your energy needs. It might be possible to eat like this if you are a recluse in your cell, praying or meditating all day in samâdhi. This *"sattvic"* activity only requires small amounts of terrestrial food. However, this so-called ideal cannot realistically be applied to everyone. It would be more accurate to say *"You should leave the table before feeling full"*. Once you have reached satiety, you have already eaten too much, which deforms the digestive system and can lead to overconsumption. However, there are people in the world who are in perfect health, but never eat, defying the laws of medicine and physiology (see the many Web sites on *inedia (the ability to live without food and water)*.

— **Eat fresh**; the most bioavailable nutrients are found in fresh, unaltered fruits and vegetables. Mechanically processed foods lose some of their vitality.

— **Eat local**; foods grown locally resonate with the physical bodies that live in the same environment. It is also an ecologically responsible choice that reduces the high energy cost of transporting foodstuffs and supports local growers.

— **Eat organic**; non-organic soil contains pesticides, insecticides and medication (antibiotics, hormones, etc.). Extremely harmful and often carcinogenic, these substances end up in the fruits and vegetables grown in that soil or in the animals that graze in the fields. Organic foods are healthiest.
— **Eat raw**; cooking alters or destroys certain nutrients, namely enzymes. If you must cook your food, cook it as briefly as possible. If you are preparing a simmered dish, at least try to know its origins to make sure that it was cooked with loving care.
— **Eat slowly**; chew slowly to give your digestive system time to secrete all the juices needed for proper digestion.
— **Eat happy**; one of the most important aspects of a healthy diet is how you feel when you sit down for a meal. Love what you eat, let it be a joy, something to savour, a treat. If you are forcing yourself to eat something you do not like, you will receive few of the benefits.
— **Eat with gratitude**; you will assimilate your food much better if you appreciate the fact that you can eat until you are full. You belong to a minority of people on this starving planet. Be grateful to the person who prepared the meal and the abundance you are enjoying. Give thanks. It is one of the fundamental reasons the different cultures make offerings, pray and say grace before meals.

If you decide to change the way you eat, be prepared for a battle on several levels. You will need to fight old habits, often anchored in your beliefs and personal, familial or societal values. Changing one's diet is often as difficult a fight as overcoming an addiction. For example:

A woman decides to adopt a strict vegetarian diet after eating meat on a daily basis. Her new diet appears to be balanced, without any theoretical deficiencies. Within a few days, she begins experiencing discomfort in several areas. She suffers from fairly intense headaches, abdominal pain, intestinal cramps, constipation, intestinal gas or diarrhea, fever, mood swings, irritation, depression or despondency... After a few days of this, she decides to quit and return to her previous diet. The symptoms disappear. She concludes that this diet was not for her. And yet...

This person had a rather typical experience, one I have seen many times in my practice. The symptoms, which vary from person to person, are exactly the same as those experienced during withdrawal! Withdrawal is no picnic, as anyone who has ever battled addiction can tell you!

Most food addictions can be cured within three weeks to one month. All of the symptoms will arise during this period, but it is important to let them come. You will emerge stronger and healthier.

You may need help, or coaching, to get through the process. You will need to do your research to find intelligent coaching based on extensive experience with healthy diets adapted to your geographical area and values. Although your doctor or nutritionist should usually provide this support, they are not always able to help you during this precise step. You will therefore need to find the expertise you need in the field of complementary and alternative medicine.

Once your body has overcome its addiction to unbalanced food, the battle is still not won. You will need to keep up your new habits and remain vigilant, because your old beliefs and resistances could crop up at any time. Or you may feel pressure from family, friends, colleagues and others to return to your old ways (and theirs), causing you to relapse.

The Importance of Hydration

As a clinical osteopath, it is my duty to provide more information about the vital role hydration plays in the physical structure. All too often, I see the ravages caused by dehydration, tied to most of the chronic problems my patients are afflicted with. Dehydration is not, as such, experienced as a feeling of thirst. "Dry mouth" is traditionally considered one of the main signs that you need to drink. However, it is an unfortunate misconception, one that is quite widespread among the various groups of therapists. The mucosa dries out because it is chronically dehydrated. Dryness is the final symptom of dehydration, not the first.

The adult physical body is made up of nearly 70% water. That percentage is even higher in pregnant women and infants. The ground substance of the connective tissue is a type of aqueous gel consisting of over 80% water. The intracellular liquid is also composed of over 60-70% water. It is therefore easy to see why regular water consumption is important.

For metabolic purposes, the adult body at rest requires 3-4 litres of water per day. Half of this can stem from fluids and water in solid food. The other half therefore needs to be supplied in liquid form. Not all liquids are equal, however. In fact, no other drink can adequately replace pure water. The cells are genetically programmed to treat water as their main source of hydration. It is true that most beverages contain water, but most also contain dehydrating agents (caffeine, refined sugar, alcohol, sweeteners, food colouring, various flavourings, etc.). These liquids stimulate micturition and are therefore dehydrating. The least harmful and most frequently suitable beverages are essentially teas (even those containing caffeine) and herbal teas with medicinal properties.

Researchers are starting to suspect that this universal molecule (H_2O) is responsible for keeping the living matrix unified and highly cohesive. Without H_2O, the connective tissues contract, shrink, wither. Aging of the tissue is therefore primarily due to dehydration.

Water is the main component of the matrix of the connective tissue. Everything is bathed in it. With sufficient hydration, the physical body's aqueous content becomes a very sensitive channel for communication between the molecules and cells of every tissue. Water acts not only as the main element binding the organism's entire connective structure, but also makes conduction possible. All of the connective tissue's molecules are polar; ionic. They carry electrical charges. And water is a semi-conductor of electricity.

This inherent communication made possible by water takes place at incomparable speed, much faster than nerve impulses.

Hydration and Gas Exchange

I will explain how important gas exchanges are for vitality in the section on breathing. However, it is worth noting that in metabolism, all gas exchanges between the blood and cells result in a state of maximal intracellular oxygen saturation (O_2). Oxygen, along with carbohydrates, is the main source of energy for the tissues.

The gases involved in these exchanges are dissolved in the aqueous environment of the fluid connective tissue that comprises the blood. Through different pressure gradients, the gases travel between the red blood cells and neighbouring cells. They can only do so efficiently if the interstitial environment is very fluid. If the interstitial liquid is too thick or viscous (dehydrated), it will not be able to adequately fulfill its role. The cells therefore do not receive enough O_2, leading to hypoxemia and metabolic acidosis, followed, chronically, by all types of pathological disorders.

Rehydration Protocol

Humans can survive more than two weeks without solid food. Some fasts can last over a month. But unless you are in a controlled state of inedia, you will not be able to survive more than three days (one week in extreme cases) without water.

If you are chronically dehydrated, it will take more than three months to adequately rehydrate all the cells of your physical structure, sometimes longer. The reason is that the body will be able to absorb larger quantities of water only gradually, with the amount increasing over time. In the beginning, you will be urinating the excess (see the paragraph on micturition below).

A normal adult at rest needs to drink approximately two litres (eight glasses) of water per day. With intense exercise, this need rises to one litre per hour or more.

A good rule of thumb is that adults at rest should ingest the equivalent of one cup of water (250 ml) for every 20 pounds (9 kg) of body weight.

Here is an example of a daily hydration protocol (for an adult weighing 62 kg or 150 pounds):

— Drink 2 cups of water upon waking.
— If you have a morning Yoga practice (sâdhana), start approximately 15-30 minutes after drinking.
— Have breakfast.
— Drink one cup of water 90 minutes after breakfast and another cup 90 minutes before lunch.
— Drink one cup of water after lunch and another cup 90 minutes before supper.
— Or, if you have an afternoon sâdhana, drink one cup of water 15-30 minutes before. Do the sâdhana; then eat your evening meal.
— Drink one cup of water 90 minutes after your meal and another 90 minutes before going to bed.
— That amounts to 8 cups of water. Adapt this protocol to your needs and weight. You can drink more if you are still thirsty.

Some authors will teach you to drink even more than that, up to 4-5 liters a day! However, please note that some researchers are now contesting all of this data, claiming that it is not based on scientifically valid studies.

Cerebral Thirst Centre

The cerebral thirst centre is located in the *hypothalamus*, deep in the centre of the brain. This centre is filled with sensitive neurons that can interpret the body's state of hydration by evaluating the electrolytic concentration of the neuronal fluids. When the electrolyte concentration increases, it means less water dilution (due to dehydration). When the neurons detect a local drop in water concentration (or a drop in the antidiuretic hormone), they send a sensory impulse that is interpreted as a thirst stimulus in the brain. The hypothalamus makes this sensory information conscious and the subject experiences thirst, which he/she should be motivated to quench by drinking. However, this nerve circuit is very fragile. It also appears to deteriorate with age (elderly people are known to be at higher risk for dehydration). Moreover, head trauma, brain surgery and concussion can affect this function. Finally, on a behavioural level, if a thirsty person does not drink, the system will malfunction. Due to a complex mechanism, the cells of the thirst centre appear to stop responding to a drop in water concentration when the body is not rehydrated. In as little as three weeks, the thirst centre can lose its ability to adequately trigger thirst. The person is therefore unaware of being dehydrated because he or she is simply no longer thirsty or, more precisely, not aware of being thirsty! It is a little as if the sensory cells had simply turned their sensitivity switch to "off".

In applying this protocol, the thirst signal will return to normal within a few days. The cells of the thirst centre will switch back to "on". This can even be quite worrying for someone who was dehydrated. After drinking two litres of water per day, when the system becomes functional again, you could be even thirstier! There is no need to worry, however. If you are thirsty, drink! If you feel thirst, quench it. You will quickly regain a healthy metabolic balance.

Micturition

Those who follow this rehydration protocol will see their daily volume of urine increase, along with the frequency of micturition. Everything should come into balance within about 3 weeks. Expect to urinate at the same frequency, but a larger volume than when you were dehydrated.

Finally, we must understand that micturition is one of the primary ways of excretion and detoxification management of the kidneys; urinating is healthy!

Salts, Electrolytes and Minerals

Ingesting such a large amount of water will lower the body's salt concentration. The kidneys will be stimulated to filter more blood to balance the acidity and water concentration of the tissues. More frequent micturition risks causing some demineralization. Usually, minerals are stored in the body as crystalline salts. In fact, the bones contain over 25% of the body's salts, which are stored as crystals and are responsible for the hardness of the bone structure, among other things. The rest of the mineral salts (sodium chloride, potassium salts, calcium salts, phosphorus, etc.) are dissolved in the various fluids and are used in metabolic exchanges as ions and electrolytes. Mineral salts or electrolytes support and meet the body's needs. Their presence helps keep the water concentration in the tissues at an adequate level. These salts are usually replenished through nutrient intake.

If, in applying this protocol, you begin to find water unappealing, experience nausea and/or headaches or are woken up by muscle cramps, your body is probably low in salts or electrolytes. Be sure to eat generous helpings of organic fruits and vegetables, which are a good source of vitamins and minerals. *You should also add a pinch of unrefined sea salt (like fleur de sel or Himalayan salt) to one of your daily glasses of water. A pinch in a single glass will meet your metabolic need for salts and minerals.* The symptoms should soon disappear.

Recognizing Dehydration

How do you know if you are dehydrated? Or if you are thirsty without being aware of it? Here are twelve signs and symptoms:

- Your urine during the day is dark yellow and sometimes has a strong odour
- Your mouth is pasty and your saliva is thick
- You constantly clear your throat and larynx
- You have rings under your eyes (maybe for as long as you can remember!)
- You always feel tired
- Your sleep is agitated and not very restful
- You suffer from frequent headaches
- You have frequent muscle cramps
- Your body fluid is more acidic
- You suddenly develop wrinkles and your skin appears to be withering
- You easily develop dermatoses
- You easily develop inflammation

Stress and Dehydration

The more of these signs and symptoms you have, the more likely it is that you are dehydrated. But before panicking about your state of health, know that this is far from a rare occurrence. Some researchers claim that over 80% of the people who consult a health care professional for a chronic problem in America (all conditions combined) are in a state of dehydration!

We can also compare this non-exhaustive list of signs and symptoms to the list for chronic stress. The clinical presentation is very similar. The two lists are practically identical, demonstrating that dehydration is a complex syndrome with many factors, in which stress appears to play a significant role.

If a person is seriously dehydrated, their blood pressure will drop, sometimes enough to cause a state of shock. This is a medical emergency. Once hospitalized, they will be treated with an intravenous rehydration solution containing sodium and other essential electrolytes.

Yoga therapists should always inquire about their client's state of hydration.

The Art of Regeneration

When you master the art of self-regeneration, solid nutrition and hydration are transformed. You learn to extract sustaining Prâna directly, with no go-between. Every action becomes a ritual, a kriyâ that disseminates Prâna in your structure. You are continually nourished, replete and satisfied. Your body, intoxicated with Prâna, is perfectly healthy, energetic and vigorous. It becomes a wonderful perceptual and creative entity. Yoga can make that happen.

This optimal evolution of the physical body is expressed at all levels, especially in the powers of perception: the senses.

The Senses

One cannot fully grasp the wonder of the living physical structure without noting that the body is a remarkable perceptual entity. The physical body has the ability to perceive its environment (internal and external) thanks to its many sensory faculties. Over 75% of the encephalon pathways (brain and cerebellum) are designed to integrate or coordinate sensory information. This information is mainly processed unconsciously. Only a very small number of the sensations that are constantly bombarding the organism reach our consciousness.

The sensory functions are mainly carried out by the sense organs. Each of these organs is a highly specific sensory receptor, usually anchored in the connective tissue and connected to the central nervous system by way of the peripheral nerves. Each receptor is an ultra-specialized sensory terminal excrescence of a sensitive nerve fibre.

Sensory information is sent at variable speeds, in the form of electrical impulses, from the periphery toward the centre (afferent impulse), where it is integrated by the spinal cord or encephalon.

However, sensory information is also instantly carried throughout the physical structure via the more or less fluid intermediary of the connective tissues and the bio-electromagnetic fields mentioned earlier. These fields are therefore closely tied with the individual's perceptual abilities.

This last point reveals the complex relationship that exists between the physical body and how it perceives its environment. However, the tantric texts delve even deeper.

Indriyas

The study of the senses in Tantra Yoga is the one of *indriyas*. It substantially broadens the study of the senses made by classical physiology. This knowledge demonstrates the subtle, in-depth understanding that yogis in medieval India had of perception.

It is not an easy task to simplify this nomenclature, but lets try it.

In the Tantras, the senses are *tattvas* (basic principles of the human experience). They arise from the mind of individuation, *Ahamkâra*, the mind of self-appropriation, the ego (part of the mind we will cover in the next chapter on the mental body). Tantrikas believe that there are ten senses; eleven if you add the lower mind (*manas*), inherent to raising the awareness of each of the senses.

Powers of Cognition

The senses are faculties or powers that inform the ego about the world. The first family of senses is made up of the five powers of cognition (*jnânendriyas*) expressed by our five classic afferent senses: sight, hearing, smell, taste and touch.

The proposed protocol for this study provides a much deeper understanding of the senses and their functions than does classical physiology. Following is the suggested study protocol for each sense:
– *Sense*
– *Physical sensory organ*
– *Element of the sense (Tanmâtra)*
– *Object of the sense*
– *Mind (Manas)*

For sight, we have:
– Sense: Sight
– Physical sensory organ: The retina of the eye and nervous pathways up to the visual cortex of the occipital lobe of the brain and their associated connective tissue links throughout the body
– Element of the sense: Light (electromagnetic spectrum, from infrared to ultraviolet wavelengths; visible chromatic spectrum)
– Object of the sense: What the eye looks at
– Mind: Mental function of integrating and interpreting the light perceived

For hearing, we have:
– Sense: Hearing
– Sensory organ: The inner ear (organ of Corti in the cochlea) and cochlear nervous pathways up to the auditory cortex of the temporal lobe of the brain and their associated connective tissue links throughout the body
– Element of the sense: Air (fluid that carries sound waves)
– Object of the sense: The origin of the sound wave
– Mind: Mental function of integrating and interpreting the sound waves perceived

For smell, we have:
– Sense: Smell
– Sensory organ: Receptors of the nasopharyngeal mucosa and corresponding nervous pathways up to the olfactory cortex of the temporal lobe of the brain and their associated connective tissue links throughout the body. The ancient texts mention the "nose"
– Element of the sense: Volatile gaseous molecule as fragrance
– Object of the sense: Source of the fragrance
– Mind: Mental function of integrating and interpreting the fragrances perceived

For taste, we have:
– Sense: Taste
– Sensory organ: Taste buds (tongue and oropharyngeal mucosa) and corresponding nervous pathways up to the gustatory cortex of the temporal lobe of the brain and the gustatory nuclei of the hypothalamus, thalamus and primary sensory cortex of the parietal lobe of the brain and their associated connective tissue links throughout the body
– Element of the sense: Volatile molecule dissolved in oral digestive fluids, producing flavour
– Object of the sense: Origin of the flavour
– Mind: Mental function of integrating and interpreting the flavours perceived

For touch, we have:
– Sense: Touch
– Sensory organ: Sensory receptors of the skin, mucosa and fascias; scattered in and around all the organs
– Element of the sense: Initial stimulus (temperature, texture, contact, movement, etc.)
– Object of the sense: Source of the stimulus

– Mind: Mental function of integrating and interpreting the sensations perceived

Powers of Conation (or Action)

The second family of senses is made up of the five powers of conation (powers of action or *karmendriyas*). These efferent faculties produce effects that heighten our knowledge of the world. They are expressed by creative activities: speech (communication), hand (manipulation), foot (locomotion), anus (digestion) and genital organs (procreation).

In our tantric study of speech, we have:

– Indriya: Speech (vocalization, verbal communication)
– Indriya organ: Laryngeal apparatus (vocal cords)
– Indriya element: Air circulating in the larynx
– Indriya object: Target of this projection
– Mind (mana): Mental function of primary integration

For the hand, we have:

– Indriya: The hand (manipulation)
– Indriya organ: Extremities of the limbs (especially the hands)
– Indriya element: Manipulated matter
– Indriya object: Nature of this matter
– Mind: Mental function of primary integration

And so on, for the other conative senses.

All indriyas could be developed in a broader way through Yoga practices within the subtle body (see section on Sûkshma Sharîra)

A word about the mind before we move on... In the tantric vision of the ontological categories, the mind is divided into three parts: the lower mind (Manas), the ego or "I" mind (Ahamkâra) and the higher mind or intellect (*Buddhi*). I will explore this idea in more depth in Chapter 2, when covering the mental body.

When a sense is activated or a conative activity is undertaken, information (afferences) begins to enter and modify a person's awareness of the world around them. The information is interpreted by the mind, which operates the central nervous system. This ensures the appropriate response to the stimulation.

Sometimes, the only response is a silent integration. The person simply knows a little more about his or her environment.

If the appropriate response is another action or movement, one or several muscles will be called upon. The majority of muscle contractions are therefore only reactions to initial sensory stimuli.

This leads us to study the phenomenon of motricity, the ability to produce contractions. By extension, this also leads to the phenomenon of locomotion, the ability to produce spatial movement. These functions are essentially attributed to the muscle structure and the locomotor system.

Muscle Physiology

Muscle
A muscle is a myofascial organ. That means that it is built as a fibrous structure made up of connective tissue containing a large quantity of *myocytes*, the contractile cells of the muscle. These cells are contractile because they contain a substantial concentration of protein fibrils with the ability to shorten by modifying their architectural morphology when stimulated by a nerve impulse.

Movement
Movement expresses life and Hatha Yoga is a metaphor for this life. The different âsanas are examples of the variety of forms through which this life can be expressed. To shift from one form the other, from one âsana to the other, the physical body must be in motion, which it does by moving the limbs and the trunk through space. These movements are possible because the body has many joints that are more or less mobile, activated by the

skeletal muscles. Contracting these muscles moves the skeleton through space. The smooth muscles, on the other hand, provide motility and movement inside the body to ensure its intrinsic metabolic function. Consequently, the skeletal muscles move the body through space while the smooth muscles move substances within the body.

But where does the body end and space begin? There is a large amount of space between the myocytes of a muscle. And there is a great deal of space between two âsanas. I often illustrate this idea to my students by saying: *"You are this space in which this âsana occurs."*

Contraction

Contraction is therefore a physiological event that occurs in the muscle cells in response to nerve stimulation.

The myocytes of the skeletal muscles are long, multinucleic cylinders. They contain many nuclei (50 to over 500) because their length (sometimes as long as the body of the muscle) requires several local nuclear controls. In these cells, contraction occurs when the protein filaments, *myofilaments*, slide within the contractile unit of the myocyte, the *sarcomere*. A sarcomere, in a myocyte, is made up of a set of protein myofilaments laid out geometrically in such a way that when nerve stimulation occurs, these myofilaments slide against one another, which shortens the sarcomere and makes it denser. This event is simultaneously reproduced in all the sarcomeres of a single cell. A skeletal muscle's myocytes can contain over one hundred thousand sarcomeres. The summation of the contractile activity of all the myocytes leads to the contraction of the entire muscle.

The myocytes of the smooth muscles do not have sarcomeres. These cells are much shorter, fusiform and mononucleic. The contractile myofilaments are laid out in a less orderly fashion. Their contraction produces a shrivelling of the cell and, by summation, the entire smooth muscle. This action is sufficient to produce the expected motility in the organ.

Contraction begins when a nerve penetrating the periphery of the muscle transmits a nerve impulse. The nerve spreads in the muscle, expanding to all the myocytes. The impulse reaches the myocyte at the synapse between the nerve and the cell membrane. The myocite part of the synapse is called the *motor end plate*. Generally, the impulse starts a cascade of biochemical events that trigger the contraction.

Muscle contraction therefore drives movement. The skeletal muscles mainly produce voluntary contraction. Involuntary contraction is mainly produced by the smooth muscles.

Performing an âsana requires first a control of the voluntary movements, therefore the action of the skeletal muscles.

Types of Muscles at Work

There are over 625 skeletal muscles around the various joints of the bony skeleton. Most of these are regulated by the somatic nervous system. Their contraction is therefore directly dependent on will.

When a skeletal muscle is in contraction, it usually produces movement around a joint.

Exercise physiologists have classified the muscles according to the role they play in producing a given movement. When a muscle is the main instigator of the movement, it is called the *agonist* of the movement.

A movement is rarely produced by just one muscle, however. Each movement is usually the result of several muscles working together. They help produce the action without being the main instigators. The muscles that support the agonist are called the *synergists* of the movement. To flex the elbow, the agonist is the biceps brachii and the synergists are the brachialis, brachioradialis, etc.

Each movement around a joint is paired with an opposite movement. The opposite of *flexion* is *extension*. At the elbow, the muscle responsible for extension is the triceps brachii. Since extension is the opposite movement of flexion, we can conclude that the triceps opposes the action of the biceps at the elbow. In fact, when the biceps induces flexion at the elbow, the triceps is stretched and its viscoelasticity produces passive resistance to the flexion. When a muscle directly opposes a movement in this way, it is called the *antagonist* of the movement. In our example, the triceps brachii is the antagonist of the flexion at the elbow. Inversely, we can say that the biceps brachii is the antagonist of the elbow extension, with the triceps brachii acting as the agonist.

In short, movement around a joint is made possible by the activity of the agonist and synergists and is subject to resistance or opposition by one or more antagonists.

Types of Muscle Contractions

Muscle contraction does not always produce the same result. There are various types of contractions for the same muscle, with different effects.

Typically, since a contraction's intrinsic metabolic process involves a linear shortening of the myocytes in the muscle belly, we should expect a linear shortening of the entire muscle. But that is not always the case.

Let us go back to the elbow flexion example. When the biceps brachii contract, the forearm approaches the humerus. Since the distal tendon of the biceps is attached to the radius, the muscle shortens as the radius gets closer to the humerus. This classic muscle contraction, in which the agonist muscle shortens, is called a *concentric* contraction.

However, it is possible for the biceps brachii to contract without producing movement. In fact, if the force generated by the contracting muscle is not sufficient to lift the weight of the limb to be moved, the movement does not take place and the contracting muscle does not shorten. A contraction in which the muscle retains its initial length is an *isometric* contraction. In this contraction, there is no movement of the limb around the joint.

Finally, it is even possible to lengthen a contracting muscle. It might seem contradictory, but this type of contraction exists and is even quite common. This is an *eccentric* contraction. In this contraction, a muscle already in the shortened position after a concentric contraction is led to hold its contraction while actively opposing its lengthening. This occurs when the exterior lengthening force exerted on the muscle is greater than the internal force of contraction (which attempts to hold the muscle in the shortened position).

In short, there are three main types of contractions: concentric, in which the muscle shortens, isometric, in which the muscle retains its initial length and eccentric, in which the muscle lengthens.

Physiology of Muscle Performance

Muscle physiology researchers have also classified the skeletal muscles according to their performance at work, that is, their ability to produce and maintain effort. They have thus identified the *myoglobin* content of the main skeletal muscles, the speed at which they contract and their resistance to fatigue.

Myoglobin is a protein similar to the haemoglobin of the red blood cells. It is found in the myocytes of the muscle. Like haemoglobin, it contains iron atoms capable of binding oxygen. A muscle rich in myoglobin can store more oxygen and use it, along with carbohydrates, to produce energy. Its energy production is therefore *aerobic*. This oxidative metabolism is possible due to a high concentration of mitochondria within the myocytes. Muscles rich in myoglobin have a reddish appearance, which is why they are called *red muscles*. These myocytes (Type 1) perform their contractions slowly and are highly resistant to fatigue. They are also called *tonic muscles*.

Muscles with the least amount of myoglobin have a whitish, slightly pinkish appearance, which is why they are called *white muscles.*

A white muscle (poor in myoglobin) cannot use a lot of oxygen to contract. It must therefore use an *anaerobic* source of energy, which it obtains by metabolizing the *creatine phosphate* molecule in its cells. This reaction releases a large amount of energy (heat), very quickly. White muscles are *phasic* muscles. Their fibres are type II and have fewer mitochondria. Their contraction is quick and powerful, but not very enduring. They tire quickly because this type of reaction produces acid waste (*lactic acid*) that builds up in the muscle and greatly diminishes the efficiency of the contractions over time. Phasic muscles must regenerate their creatine phosphate and get rid of their acid waste before being functional again. In muscle physiology, this is called the post-effort recovery phase.

Physiologist Janda's identification and clarification of how the tonic and phasic muscles operate is now classic:

- The main tonic muscles are activated in flexion synergies.
- The phasic muscles are usually the antagonists of the tonic muscles, activated in extension synergies.
- Tonic muscles functionally perform concentric contractions and have a tendency to shorten with stress. They mainly benefit from stretching.
- Phasic muscles functionally perform eccentric contractions and have more of a tendency to weaken when lengthening. They mainly benefit from strengthening.

Most of the muscles contain the two types of fibres in variable proportions. Physical training can affect these proportions and therefore increase performance. In the second part of the book, I will explain how Hatha Yoga potentiates this knowledge.

Cocontraction

Cocontraction (or *coactivation*) is a muscular event in which the antagonists contract simultaneously. If the antagonist muscles around a joint contract at the same time, with similar force, the net result will be no movement, since the flexor and extensor cancel each other out. The muscles are therefore in simultaneous isometric contraction.

In an âsana, the cocontraction of the antagonists is only fully possible in the established pose. It will be dampened during the movement toward the final pose (*vinyâsa*), because another physiological phenomenon, *reciprocal inhibition*, comes into play: (see next the section).

Cocontraction is not solely the domain of a joint's antagonists. In fact, since all of the muscles are part of a myofascial whole, when two or more muscles are in simultaneous contraction in an âsana (or in another movement or pose), we can generally consider them to be in a state of cocontraction. In short, we are functionally in a constant state of cocontraction.

Even though it essentially involves the skeletal muscles, cocontraction is not usually a matter of will. The poses are held via a more or less pronounced cocontraction that is part of the postural automatisms acquired by repeating movements during the growth period, which remain under the governance and neuronal coordination of the cerebellum.

Cocontraction becomes a therapeutic tool mainly when it involves the antagonists around a given joint and consciousness voluntarily takes over its control.

Reciprocal Inhibition (reciprocal innervation)

Reciprocal inhibition is a neuromuscular activity that occurs with every movement. It is initiated in the spinal cord. When a motor nerve impulse is created in the neuron and sent through the nerve to a given muscle, another nerve impulse, the inhibitor, is simultaneously sent to the antagonist muscle through a branch of the same nerve. As such, when the motor impulse arrives at the biceps brachii to produce flexion, an inhibitor impulse impedes the antagonist's contraction (triceps brachii); this is *reciprocal inhibition*. Resistance to flexion is therefore greatly diminished. This physiological event not only favours agonist movement, but also provides better coordination of the movement.

However, it is important to understand that this inhibition is not strong and it is possible to voluntarily conteract the antagonist even when the agonist is engaged. Reciprocal inhibition is an automatic function and easily yields to voluntary activity. When the biceps brachii is contracting, the triceps is inhibited, but not completely. It can be contracted during bicipital contraction. It is therefore a cocontraction in which the biceps overcomes the contraction of the triceps. In this case, the agonist is acting concentrically and the antagonist is acting eccentrically.

Using reciprocal inhibition can help further lengthen the antagonists of the poses when performing âsanas.

Later, we will see that conscious cocontraction is likely the safest means of exploring movement in most physical activities, including Hatha Yoga of course, as well as providing support during recovery from musculoskeletal injuries.

According to exercise physiology theories, regular exercise increases the muscles' ability to store oxygen, resulting in greater endurance and overall performance

The beauty of Hatha Yoga, at this level, is that a regular practice focusing on âsanas and prânâyâmas increases the aerobic capacity of the cardiovascular and muscle systems. When done correctly, it is probably the physical training activity with the least impact and risk of injury.

Abnormally high stress, negative emotions and traumatic accidents can all cause dysfunctions in this myofascial system. The muscles then enter into continuous protective contraction, causing all manners of problems, symptoms and pain. This is even more serious when it occurs at the axial level. In the central axis, it is not uncommon to see red muscles in a state of contraction for weeks, even months and years! We will see how to manage these conditions in the second part of the book.

Axial Cocontraction

The axial muscles of the trunk surround the spinal column. They are found all along its length, in the front and back. At the scapular and pelvic girdles, there are muscles on each side of the cervical and lumbar spine, connecting these vertebral levels to the head, rib cage and pelvis.

All of these axial muscles serve as active braces, holding the trunk (and especially the spinal column) in its erect posture. Their simultaneous contraction causes a cocontraction, creating a dynamic balance that counters the gravitational pull. Posture is held against the force of gravity due to a complex neuromuscular automatism involving coordination between the plantar and articular sensory receptors, the brain (*vestibular pathways*), the cerebellum and the postural muscles.

Diaphragms

At the lower end of the head, the upper and lower ends of the rib cage and the pelvic floor, the myofascial tissues form curved horizontal structures (like a dome) that connect the anterior, posterior and lateral walls of the central axis. This myofascial tissue acts like a *diaphragm*.

In the anatomical position, a diaphragm is a dynamic myofascial structure, built horizontally, that provides a transition between two geographical areas of the body. By definition, the following family of tissues are considered diaphragms (from top to bottom): *tentorium cerebelli, the tongue and floor of the mouth, the scalenes and-pleural dome, the thoracic diaphragm, pelvic floor and even the plantar fascia.*

All of these diaphragms are a part of the body's axial and postural muscles. The thoracic diaphragm is the most well known. It is one of the most important because it plays a key role in pulmonary inspiration, among other tasks. It is also—and this is not as well known—the main venous return pump supplying the right heart.

The diaphragms form all of the myofascial mechanisms needed for the body's health and energetic fluidity. Hatha Yoga and the prânâyâmas help release the tension of the diaphragms and restore even more health and power to the human structure.

Cocontracting certain axial muscles near the diaphragms creates a *Bandha.*

I will identify these bandhas and their use in the second chapter.

Viscoelasticity, Suppleness, Laxity and Flexibility

The muscles and their fascial environment must be able to contract without impediment. They must be able to freely shorten, lengthen and return to their resting length. A muscle's ability to do so is called *viscoelasticity.* It is the myofilaments that must slide smoothly in the myocytes' sarcomeres. This fluidity is gradually developed with training. Regular repetition of the movement releases heat around the myofilaments and diminishes the friction and resistance of the connective tissue in and around the cells. This results in improved muscle response and therefore greater viscoelasticity. In these conditions, the muscle is said to be *supple.* It responds well when required to stretch and return to its resting length.

A muscle is considered supple when its myocytes can contract freely, in perfect viscoelastic compliance. This is only possible when the structural connective tissue of the muscle around the myocytes is free to move. Its collagen fibres must be free and loose.

Laxity applies mainly to the connective tissue of the ligaments and articular capsules. This term is synonymous with a state of released tension, of suppleness in the very dense collagen fibres at this level.

The muscles and ligaments are considered *flexible* when they are strong and supple.

Strength and Suppleness Training
– Requires patience and intention
– Strength is needed to develop suppleness
– Suppleness is needed to develop strength
– Both strength and suppleness develop in the presence of heat
– Regular (daily) practice is required

To develop strength and suppleness, we saw that the connective tissues around the muscles must be free of impediments or tension. As we have learned, in addition to the exercises, a healthy diet and adequate hydration are vital to achieving this goal.

Power

The strength capacity deployed by the neuromusculoskeletal system is limited to the physical properties of the materials our bodies are made of. However, true strength, with a capital *S*, glorified strength, is Power.

Power is a complex notion referring at once to philosophy, physical science and the individual. Individuals can be powerful beyond their simple physical structure when their three bodies or five sheaths are in fluid and stable harmony. This is another way to express the aim of yogic practice.

Everything that has just been mentioned can only be accomplished with proper breathing. Now it is time to address the phenomenon of breathing I touched upon earlier.

Functional Anatomy of Respiration

The respiratory function extracts oxygen (O_2) from the immediate environment and uses it as one of the main sources of metabolic energy. For the accomplished yogi, respiration is the chief means of extracting Prâna, the life force I will cover in greater detail in Chapter 2. Respiration also helps the body use *carbon dioxide* (CO_2), cellular metabolism's main gaseous product. It essentially consists of two separate, but closely related physiological activities: *pulmonary respiration* and *cellular respiration.*

The physiology of respiration is the scientific study of this function. To date, science appears to have gained a good understanding of the biomechanics of respiration and the biochemistry of gas exchanges. Let us take a closer look at these phenomena.

Pulmonary Respiration

Pulmonary respiration causes ambient air to move through the bronchopulmonary tree. This circulation is under the nervous control of the respiratory centre located in the medulla oblongata of the central nervous system. This centre controls the biomechanics of the respiratory muscles. Inhalation (*puraka* in Sanskrit) moves air from the nostrils to the pulmonary alveoli. Exhalation (*recaka*) moves the inhaled air in the opposite direction.

A higher concentration of carbon dioxide (CO_2) in the blood will stimulate the cells of the medulla oblongata respiratory centre. The chemoreceptors in charge of this sensitivity are located in the carotid artery and the aorta. When activated, these cells command the respiratory muscles into action. Usually, this is an automatic rhythmic activity, a function of the autonomic nervous system.

Thoracic Diaphragm

The main biomechanical structure responsible for pulmonary respiration is the *diaphragm* muscle. It is roughly the shape of a parachute, deployed between the thoracic and abdominal cavities.

Its upper middle portion is not attached and rises within the thorax. Its peripheral and lower border is attached around the lower rib cage, from the front to the back, all the way to the spinal column.

This muscle is a fascinating myofascial structure. Its unique anatomy gives it movement capabilities found nowhere else in the human body. It is one of the rare skeletal muscles to not act directly on a skeletal joint. Its action is essentially pulmonary.

But because of its unique geographic situation, sandwiched between the human body's main vital systems, the diaphragm plays indirect, though important roles in thoracoabdominal nerve regulation, blood and lymphatic circulation and digestion.

The peripheral tendinous attachments that bind it to the skeleton are, from front to the back:

- The intrathoracic face of the xiphoid process and the 7th to 12th costal cartilages.
- At the spinal column, a tendinous pillar descends over the vertebral bodies, on both sides of the median line, the *pillars of the diaphragm* (crus diaphragmi); from L_1 to L_4 on the right side and L_1 to L_3 on the left side.
- The lateral part of each pillar rises to form the *medial arcuate ligament*, ending on the tip of L_1's transverse process.
- From the tip of L_1, the diaphragm forms another ligament: the *lateral arcuate ligament* that reaches the point of the 12th rib.
- The medial part of each pillar goes up the column and, upon reaching the opposite pillar, forms (in the middle of the spinal column) another arch directly in front of the intervertebral disc between T_{12} and L_1: the *median arcuate ligament*. This ligament makes up the *aortic hiatus*, the aperture through which we find the thoracic aorta and the *Cisterna chyli*, the start of the *thoracic duct of the lymphatic system*.

The upper part of the diaphragm is dome-shaped, convex at the top, the summit of the parachute. The surface is laterally bilobated in two cupolas with a slight sag in the middle. The cupola on the right is slightly higher than the one on the left, in the thorax, likely due to the presence of the liver just beneath. This freestanding dome is not attached to any bone. It has a thin, but solid aponeurotic central region, the *central tendon* of the diaphragm, whose fibres are dense and very closely knit.

The central tendon does not contain any myocytes and has a trefoil shape (3 leaves). The right leaf, which is longer than the left, supports the pleura and the right lung. The left leaf supports the pleura and the left lung. The central leaf, the largest, extends forward, directly behind the xiphoid process and supports the pericardium. The heart therefore literally sits on the diaphragm. The upper attachment of the *peritoneum*, the serous sheath of the abdominal organs, is affixed under the central tendon of the diaphragm.

The (contractile) muscle portion of the diaphragm runs down from the central tendon. When the person is standing, the muscle fibres of the diaphragm are essentially vertical, curving horizontally over the dome in the direction of the central tendon.

To contract, the diaphragm needs to be stimulated by innervation. The nerve that innervates the diaphragm is called the *phrenic nerve*. This nerve originates from the cervical region, with its roots emerging between C_3-C_4-C_5. The two phrenic nerves (left and right) travel down through the neck, in front of the anterior scalene muscle and plunge into the thoracic outlet between the subclavian artery behind and the subclavian vein in front. In the upper part of the mediastinum, it descends into the thorax between the pleura and pericardium. It sends sensory fibres to the pericardium and the diaphragm. And it sends its motor fibres on the diaphragm over which it spreads to its upper and lower surfaces.

Foramen and Relations of the Diaphragm

In a frontal cross-section of the abdominal trunk, we can see that the concave lower surface of the diaphragm relates on the right to the right lobe of the liver, right upper kidney and right adrenal gland. It relates to the left to the left lobe of the liver, fundus of the stomach, upper left kidney and left adrenal gland.

Important organs cross the diaphragm, connecting the thoracic cavity to the abdominal cavity. The diaphragm has three major apertures or openings. I spoke earlier of the median aperture between the two pillars. That is the aortic hiatus making way for the *thoracic aorta*, which, at that point, becomes the *abdominal aorta*. The abdominal aorta branches into an important arterial trunk, the *celiac trunk*, at this precise location. The celiac trunk supplies blood to the liver, gallbladder, stomach, pancreas and spleen in the abdomen, among others.

Directly above the aortic hiatus, the diaphragm's right pillar, which rises into the posterior muscular portion, folds back on itself and forms a loop of muscle: the *esophageal hiatus*. The esophagus passes through this aperture before becoming the stomach. In this foramen, the esophagus is accompanied and surrounded by the plexus of the two *vagus nerves*. The vagus nerves (10th pair of cranial nerves) originate in the brainstem and are the main source of *parasympathetic* autonomic nerve fibres in the body. From there, the vagus nerves wrap around the celiac trunk in front of the aorta and, by anastomosing with the fibres of the *sympathetic* autonomic nervous system, form the *celiac plexus*, making up the main part of the solar plexus in the upper abdomen. It is starting from this plexus that all the nerves of the autonomic nervous system run toward the abdominal organs.

The diaphragm's largest aperture is located where the right leaf and central leaf of the central tendon meet, called the *vena caval foramen*. This foramen is completely fibrous, oval-shaped and, in adults, has a long axis that is 3-4 cm in diameter. The inferior vena cava passes through this aperture, returning venous (non-oxygenated) blood from the abdomen to the right atrium of the heart. The *adventitia* of the vena cava (its fibrous peripheral connective tissue layer) is securely attached to the fibrous edge of the foramen.

Venous Return, Lymphatic Return and Diaphragm

It is important to understand the rather unique role the diaphragm plays in venous return to the right heart. While pressure in the peripheral arterial network ensures relatively smooth blood flow in all of the body's arteries and arterioles, pressure in the peripheral venous network is almost non-existent. Venous blood must return to the right heart against the force of gravity, in veins whose muscles cannot ensure adequate return by themselves. The skeletal muscles next to the veins therefore assist venous return. The muscles compress the venous walls by contracting, helping drive blood to the heart. But even that effort is limited. This is where the diaphragm comes in.

During diaphragmatic contraction, the dome descends toward the abdomen. This biomechanical action not only causes a drop in pressure in the lungs, but also a rise in pressure in the inferior vena cava. On exhalation, the diaphragm releases its contraction and rises elastically in the thorax. The thin valve between the vena cava and the right atrium yield to the strong abdominal pressure exerted upward on the inferior vena cava. The venous blood is then pulled into the right atrium of the heart via a sort of suction effect. The diaphragm's role as a venus pump is still not well known, even though it is probably the most significant mechanical contribution to venous return. If the pump for systemic arterial circulation is the left heart, the pump for systemic venous circulation—its heart— is the diaphragm!

As for lymphatic return, over 60% of the human body's lymph nodes are concentrated around the sub-diaphragmatic organs and most empty into the thoracic duct's cisterna chyli, the largest lymphatic reservoir of the human body. Due to its position, this duct is continually massaged by the contractions of the aortic pulse and experiences the same pressure from the diaphragm. Its contents are thus efficiently pumped toward the heart.

Structure and Biomechanics of Pulmonary Respiration

Pulmonary respiration is a cyclical activity attributed to changes in pressure between the outer atmosphere and the intrathoracic cavity. There are essentially two phases: the *inspiratory phase* and *expiratory phase.*

The inspiratory phase is characterized by a drop in intrathoracic pressure caused by a contraction of the thoracic diaphragm. When contracting, the diaphragmatic dome lowers, dragging the lower pleura and lungs as well as the pericardium with the heart toward the abdomen. This causes an increase in intrapulmonary volume, which in turn lowers intrapulmonary air pressure. Since the atmospheric pressure of the air becomes higher than in the lungs, the intrapulmonary depression causes air to move into the lungs to balance the pressure (or fill the void). The air enters the lungs until the atmospheric and intrapulmonary pressures are equal. In respiration at rest, the input of air is about 500 millilitres. In pneumology, this is called *tidal volume* (TV). During the inspiratory phase, the intrinsic elasticity of the lungs and costal cartilages create an accumulation of potential kinetic energy in these structures. When diaphragmatic contraction ceases, at the end of inspiration, this energy is released and the ribs and pulmonary tissue resume their initial position. This is done without effort or expenditure of energy since the energy was already available in the wake of inspiratory accumulation. When returning to its position at rest, the rib cage compresses the inflated lungs and increases intrapulmonary pressure. This causes air to move out of the lungs because the intrapulmonary pressure is higher than the atmospheric pressure. That is exhalation. The expiratory phase takes place as long as the atmospheric and intrapulmonary pressures are not balanced.

Forced Respiration

The physiology for pulmonary respiration in a person who is not at rest is completely different. Where pulmonary expiration at rest is passive, all other forms of diaphragmatic activity require energy expenditure.

At maximum inspiration, the diaphragmatic dome quickly reaches the limit of its descent. It cannot drop more than 1 cm into the abdomen because the central traction exerted by the heart vessels that penetrate the hilum of the lungs resists the pull. After 1 cm, the central tendon becomes a fixed point. If the diaphragmatic muscle continues to contract, its lower section, the costal attachment, becomes mobile, pulled upward from its fixed point at the central tendon. The ribs then start to lift. The costal ascent causes an increase in intrathoracic and intrapulmonary volume that permits a larger inflow of air. This is called forced inhalation.

The process is inverted during forced exhalation.

There are hundreds of ways to breathe, varying rhythm, frequency and range of respiratory cycles. Each type of breathing immediately causes the body's other systems to adapt.

Relationships between the Diaphragms

We have seen that respiration, activated by the thoracic diaphragm, takes place in the lungs due to an interplay of pressure. During inhalation, pressure drops in the thorax and rises in the abdomen. This increase in pressure is felt all the way to the diaphragm of the pelvic floor (essentially made up of the *levator ani* muscle), pushing it down. The pelvic diaphragm rises again during thoracic exhalation.

There is therefore a synchronicity between the thoracic and pelvic diaphragms during pulmonary respiration. This synchronicity, though subtler, is also seen in all of the other diaphragms mentioned earlier: the tentorium cerebelli, the tongue and floor of the mouth, the scalene and pleural dome and the plantar fascia.

Cellular Respiration

Cellular respiration is the exchange of gases between the alveoli, capillaries, their erythrocytes and the organism's cells. These exchanges mainly involve oxygen (O_2) and carbon dioxide (CO_2).

When inhaled air arrives in the pulmonary alveoli, the alveolar oxygen diffuses to the hemoglobin of the pulmonary capillaries' erythrocytes. In return, the red blood cells transfer their carbon dioxyde content to the alveoli. This CO_2 is then expelled into the atmosphere during the next exhalation phase. Only part of the exhaled gas remains in the upper airways at the end of exhalation. This is known as dead space. At the next inhalation, the CO_2 in the dead space will be inhaled again with newly arrived oxygen.

The oxygenated blood leaves the lungs by way of the pulmonary veins, travelling toward the left heart. This is called *lesser circulation*. The left ventricle propels the blood to the capillaries of the tissues via the systemic arterial system. This is called *greater circulation*.

When the organism is in optimal health, the O_2, in the erythrocytes of the tissues' capillaries diffuse freely to the cells according to the pressure gradients present. The cellular CO_2 is diffused to the capillary blood and begins the systemic venous return trip to the right heart and pulmonary (or lesser) circulation. The cardio-respiratory circuit is thus complete.

The main metabolic goal of respiratory circulation is to reach maximal O_2 saturation in the tissues (98-99%). Human beings have few oxygen reserves (this gas is 20 times less soluble than CO_2). At rest, at sea level, we only have enough to survive 4 minutes, after which we quickly enter oxygen debt and hypoxemia. Maximal saturation therefore ensures that the oxygen is used efficiently as a main source of metabolic energy.

The entire art of Prânâyâma Yoga, in relation to the physical sheath, is to establish an effective respiratory framework to achieve this saturation.

Clearing Up Misconceptions about Respiration

On my yogic path, I have frequently come across simplistic statements about the physiological effects of respiration, oxygen and carbon dioxyde. I would like to debunk some of the misconceptions that still exist and are spread, often due to lack of knowledge, by yogis and even Yoga therapists.

- *Deep breathing is vital to proper oxygenation.* That is not quite true and needs to be understood. I have often noted that Hatha Yoga teachers lack knowledge of the physiological mechanisms involved in efficient respiration. Efficient respiration saturates the organism's cells with oxygen. Thankfully, the majority of teachers promote diaphragmatic respiration, which is effectively the best means of ensuring maximal saturation. However, few people really know why this respiration is more efficient than hyperventilation, for example. Watch the breathing of a healthy person at rest. You will first note a very slight flow with very little movement of the stomach and no thoracic movement at all. It is scientifically proven that this shallow resting respiration has one of the highest oxygen saturation rates (98-99%). It is therefore not the depth of the respiration that makes it efficient, but its slowness and calmness. If respiration is deep, but calm, oxygen saturation will occur. Anyone teaching efficient respiration should insist on diaphragmatic respiration (predominantly abdominal). Fully ventilating the lungs results in better oxygenation.

Oxygenated air reaches all the way to the bottom of the lungs and releases a proportionally larger amount of O_2 to the red blood cells than the upper parts of the lung (more involved during upper thoracic respiration).

– *Hyperventilation results in better tissue oxygenation.* Again, this is false. First, we need to agree on an adequate definition. Hyperventilation is a state in which the individual breathes a larger quantity of air per minute than the usual volume at rest. In a healthy individual, the norm (normal respiratory minute volume) is around 6 000 ml/min. That is about 12 respirations/minute multiplied by the tidal volume of air inspired at each breath (500 ml.). When an individual breathes continually at a frequency of more than 18-20 respirations per minute (frequently seen in patients suffering from chronic illness as well as in seemingly healthy people who breathe from the upper thorax), the respiratory minute volume can easily exceed 8 or 10 L/min. When that happens, carbon dioxyde is expelled in larger quantities than it is inhaled. This state in the lungs is called *hypocapnia*. CO_2 is absolutely vital for good diffusion of arterial O_2 from the hemoglobin to the tissues. Without CO_2, in hypocapnia, the hemoglobin keeps its O_2 and the tissues enter oxygen deficiency, or *hypoxemia*. Consequently, there is a large amount of oxygen in the blood, but it is not released to the cells. It is the presence of CO_2 in the arterial blood that causes a drop in blood pH, triggering acidity, which stimulates the release of O_2 toward the tissue (this biochemical sequence is called the *Bohr Effect*. In short, hyperventilation brings a large amount of oxygen to the lungs, but little to the tissues.

– *Carbon dioxyde is a toxic waste product that needs to be eliminated.* False! CO_2 is not a waste product. It is the main gaseous product of cellular metabolism. Cells need energy for their metabolic activities. They use O_2 with carbohydrates in the mitochondria to release the energy they need. This process produces water and CO_2. Normally, carbon dioxide enters the red blood cell of the cell's capillary, in exchange for oxygen. The CO_2 is then transported to the right heart and the pulmonary capillaries, where it diffuses in gaseous form in the alveoli in order to be exhaled into the atmosphere. This process maintains the blood's acid balance (pH). But not all of the CO_2 is exhaled. Some always remains in the dead space of the upper airways at the end of an exhalation and is reused during the next inhalation. The presence of carbon dioxide in the red blood cells of the tissues' capillaries momentarily lowers the blood pH. This loosens the binds between the hemoglobin and oxygen and facilitates the release of O_2 to the cells. Moreover, the presence of CO_2 in the blood causes the arterioles to vasodilate, which brings still more O_2 to the tissues' cells. Carbon dioxide is, in fact, the most powerful endogenous vasodilator and also one of the most powerful bronchodilators. In other words, carbon dioxide is vital for maximum oxygen saturation. For this reason, we should not attempt to eliminate it at all costs!

Yogâsana Practice

The end of this chapter covers the practice of yogâsanas. Having understood the essence of how the physical structure operates, we use these poses to maintain smooth and harmonious form and function. Hatha Yoga poses are also wonderful tools for transcendence, helping guide the practitioner to stunning realizations. They can either maintain your current state or propel you even further in your evolution as a human being.

Yogâsanas are very powerful. If they are misinterpreted or not well understood (*avidyâ*), they can cause imbalance and injury. Âsanas whose essence has been grasped can become, in themselves, very effective therapeutic tools for the physical body. I will explore this in the second part.

Consciousness of Yogâsanas

For study purposes, we can break each âsana practice down into three parts or phases:

> **The first is the Active Dynamic (AD) phase:**
> The active part of a pose's practice is often called the *vinyâsa* phase. In this first phase, the practitioner performs all the positions necessary to reach the sought-after pose (each of these positions is itself an âsana).
>
> **The second is the Passive Dynamic (PD) phase:**
> This phase begins after settling into the sought-after pose. The practitioner breathes freely and deeply while visualizing and feeling the flow of Prâna characteristic of this unique yogâsana. The passive dynamic phase can be held for several seconds or respiratory cycles or as long as desired.
>
> **The third is the Passive Resolution (PR) phase:**
> In this last phase, the body and mind relax, releasing all tension, in order to integrate the benefits of this âsana. This is often done in the corpse pose or *savâsana*; but it is also possible to adapt this moment of integration and relaxation to standing or sitting poses. It is usually during this phase that one truly feels the benefits of the practice. Here endorphins are released into the bloodstream.

Effectively performing any âsana always involves the same components. Let us see how this applies to each phase.

Active Dynamic Phase

- Dhâranâ: The mind is highly focused, concentrated on performing each movement.
- Prânâyâma: Each movement is guided by soft, even breaths.

- Conscious reciprocal inhibition: A continual readjustment and re-evaluation of the pose's alignment results in a wider range of motion (stretch).
- Cocontraction: Dampened during this phase, cocontraction still allows for powerful action (strengthening). The bandhas are usually introduced here.

Passive Dynamic Phase
- Dhâranâ: The mind is highly focused, concentrated on holding the pose.
- Prânâyâma: Each breath is soft and even.
- Cocontraction: Isometric contraction of the main muscles needed to hold the pose. The bandhas are maintained. The rest of the body can continue to relax.
- Surrender: Slow and gentle stretch while breathing or yielding to gravity.

Passive Resolution Phase
- Dhâranâ: The mind is highly focused on relaxation, dissociation, vacuity.
- Prânâyâma: Each respiration remains soft and even.
- The body is at complete rest: Often in savâsana.

Even if Yoga is with you at all times, it is often here that Yoga can arise...

Secrets of the Consciousness of Yogâsanas
- The three phases of an âsana are not linear.
- Each phase contains the two others.
- While practicing a yogâsana, you can enter one phase, leave it to enter the next or freely return to the previous one.
- This process is the inexorable, necessary and constant holographic exploration of an âsana.
- Yoga is NOT a linear process.

- In most yogâsanas, part of the body can very well be maintained in the AD or PD phase while the rest of the body is
  already in the PR phase.
- This interplay between the three phases will vary from person
  to person.
- Adaptation and compensation are necessary to reap the
  benefits (*siddhis*) of the âsana.

Through this knowledge of the body and the way it operates
and through the recognition of the path taken in this exploration,
we are building a bridge toward a greater understanding of the
essence and meaning of this incarnate presence.

With this foundation laid, we can move onto the second chapter, which explores the subtle dimension of the Being.

CHAPTER II

SÛKSHMA SHARÎRA
THE SUBTLE BODY

The Three Sheaths
of Sûkshma Sharîra

- The vital sheath   Prâna-Maya Kosha
- The mental sheath   Mano-Maya Kosha
- The intellectual sheath   Vijnâna-Maya Kosha

General Overview

The subtle body is the subject of much debate in yogic circles. My intention is not to convince or sow the seeds of doubt, but simply to try to clarify a concept that holds little interest for a large number of thinkers, due to its immateriality. I will open with this: what if the subtle body was fact? I decided to subscribe to this theory and strive to better define it, up to and including its potential therapeutic benefits.

This second chapter, Sûkshma Sharîra, is divided into three parts, one for each sheath. The first will explore *Prâna-Maya Kosha*. Here, I will introduce the concepts of Prâna and its role in the physical structure touching the vital sheath.

I will also cover the key connections between Prâna and Hatha Yoga and describe ways to influence prânic fluidity, for example, with bandhas. I will then introduce the basic notions of Prânâyâma and present some of the prânâyâma's therapeutic functions.

The prânâyâmas will lead into the second part of the chapter, on the mental sheath, the mind or *Mano-Maya Kosha*. In this part, I will talk about *manas* and the mechanisms that produce thoughts and ideas.

I will also explain how the behaviours of our physical and mental structures are linked and unconsciously obey the orders given by the mind. I will explore how the brain and most of the physiological functions are directly related and dependent on our beliefs. I will follow this with an explanation of the difference between our lesser and greater beliefs, which cause us to discriminate and make distinctions. This will lead into the third part, an examination of the intellectual sheath, *Vijnâna-Maya Kosha*.

In this last section, I will introduce the concepts of ego (or *Ahamkâra*), reflection, concentration, introspection and meditation and examine their therapeutic potential.

The Vital Sheath - Prâna-Maya Kosha

Overview

This sheath of vital energy is the subtle part of the being, responsible for creating and maintaining the living structure. It is what makes it possible for the physical and mental body to be alive. It is called "subtle" because it is not material. It has a higher vibrational frequency that allows it to pass through, inundate and influence matter. The body's cells and tissues have a natural affinity for the frequencies of these vibrations. They are in harmonic resonance with Prâna-Maya Kosha. The physical body can therefore express its human specificity thanks to this primordial breath that inhabits and animates it. Prâna-Maya Kosha is also the vital link between the body and mind.

Yogic theory extensively covers the concepts that follow. They are well documented and now form a complete paradigm of knowledge. I will be touching on *prâna, nâdîs, cakras,* etc. I will then summarize these notions and add a few original touches.

Prâna, Science and Tantra

Prâna is to India what *Qi* is to China, *Ki* to Japan, *Mâna* to Hawaii or the *Holy Spirit* to Christianity. In every culture, through all the ages, sages and masters have identified it and claimed it to be the causal substance of life. It appears in the Rig-Veda, the oldest of the Vedas. The Atharva-Veda, written long before the Tantras, connects it to the notion of vital breath.

In Tantra Yoga, Prâna is the fundamental element of life; life force, vital energy, breath of life. This last term is especially appropriate because it evokes the image of a "wind", a "breeze" that sows life wherever it blows.

Shakti blows this wind: cosmic energy, universal energy, the active expression of Shiva unmanifested or unchanging Brahman,

Allah to Muslims, God the Creator to Christians. Shakti, the manifested half of Shiva, is the universe's active energy. Everything originates from Shakti and is created by Her. Her creation is Nature. Tantra is nothing more than an ongoing celebration of Shakti. When Shakti creates life, Her creative impulse is called Prâna. Prâna not only allows life, but literally creates the body-mind.

Since the beginning of time, this concept has linked two notions: circulating vital energy (or Prâna) and respiration. In fact, there is a fascinating etymological connection between the two terms. The word *respiration* (re-spiration) has Latin roots related to *spiritus* (spirit), which stems from the verb *spirare* (blow, breathe). The word *spirit* (and its derivatives *inspiration* and *expiration*) is related to the Indo-European root *praan* (blow, breathe), which gives us Prâna. Not only does Prâna convey the meaning of breath, but also spirit, the spiritual breath. Therefore, India's sages made the connection between respiration and the spiritual experience a long time ago. It is another facet that justifies the obvious energetic relationship between the practice of prânâyâma and its direct effect on the inner self.

Prâna's iconography often demonstrates its fluid nature, such as a river of light flowing through the body's structure.

The yogic observation of Prâna-Maya Kosha appears to be accessible to individuals who are prepared to dedicate themselves through various means of concentration and meditation. Temples, ashrams and monasteries are filled with individuals, men and women, who have devoted their life to this study.

Prâna, Earth and Ecology

Prânashakti is everywhere on Earth. Seen from space, our planet looks like a magnificent blue ball, unique in the solar system. Prâna's basic chromatic vibration is exactly the same intense blue. The atmosphere surrounding Earth is Prâna's planetary container. This understanding forms the basis of the concepts found in various cultures and philosophies concerning the planet's vitality. The idea that Gaia (see J. Lovelock on this subject) is alive is the mystical basis of our union with Nature and the philosophical source of a good number of the ecological movements that have sprung up among the planet's young people since the 20th century

Connection with Anna-Maya Kosha

Prâna-Maya Kosha is the vital body linking the physical sheath, Anna-Maya Kosha, to the mental sheath, Mano-Maya Kosha and the intellectual sheath, Vijnâna-Maya Kosha. All of these sheaths originate from Prânashakti, the fundamental breath of cosmic energy, represented in humans by its most subtle form, *Ânanda-*Maya Kosha, the body of bliss, home of the soul.

Prâna is fluid by nature. It circulates. It is therefore an essentially dynamic substance. Its movement and fluidity are vital to its effective manifestation in matter. When a human has great vitality, radiant health and infectious energy, his or her Prâna is flowing freely and is in harmonic resonance with the tissues. When a human is experiencing pain, suffering and illness, his or her Prâna is restricted, its fluidity disrupted.

In the first chapter, we saw that the physical sheath has a framework made of connective tissues and that this tissue can take several forms, from the most dense and rigid (bone) to the most subtle and flowing (bodily fluids). This connective morphology is continuous, from the tips of the toes to the roots of the hair, forming a living matrix, an uninterrupted network of channels containing various vibration and circulation levels. All of these circulations produce as many bioelectromagnetic fields. Their summation could explain, in part, the subtle emanations of the living body.

It is said that the vital sheath harnesses the bioelectromagnetic and vibratory energy of this entire connective network's metabolic activity. The paradox is that the network of connective tissue also originates from this vital sheath. Shakti is the Creator of the Universe while Prânashakti is the maker of living matter, the physical and mental sheaths in humans.

Hypothesis

Prâna would therefore be the vibrating Light that, when the spermatazoid fertilizes the egg, emerges as the initial cell, the embryonic zygote. From there, all ontogenic development of the human being is only a temporal interlude, a divine game, the dance of Shiva, which is none other than the irresistible development of the return to Oneness.

Nâdîs and Cakras

Prâna-Maya Kosha, the vital sheath, overlays the physical structure exactly like fabric draped over the human form, in such a way that the entire physical sheath is flooded in Prâna.

Inside the body, this fabric spread in all directions. Every organ has its prânic double throughout every layer and in every area of the physical structure. We can therefore refer to the prânic liver, prânic uterus, prânic biceps, etc.

To visualize the vital sheath, picture the fabric. It has a framework whose woven fibres follow the flow of the canvas. This canvas determines the direction favoured by the fibres, which gives the whole its general aspect, its particularity. When you look at the fabric, your eyes follow the lines woven on the canvas. In the vital sheath, Prâna flows everywhere, but has a tendency to condense in preferred channels. These channels are energy pathways called *nâdîs*. According to yogic theory, there are over 72 000 nâdîs. This is probably more a figure of speech than an exact number, to illustrate the sheer volume. What is important is that Prâna, by circulating, forms the vital current called nâdî. The nâdî concept is similar to the idea of meridians in Chinese medicine.

When the nâdîs converge and intertwine, they form energy wells or vortices. A parallel can be drawn between these energy intersections and the acupuncture points. The main vortices are found along the central axis. These are *cakras* (pronounced *tshakras),* or wheels, or *padmas* (lotus flowers).

Depending on the school of thought, there are approximately six cakras along the subtle spinal column and one at the top of the skull. Other schools claim four, nine or even twelve cakras. Each cakra is tied to an element of Nature (*bhûtas*), a sensory modality, an action organ, a fundamental sound (*bija mantra*) and various divine attributes, usually expressed by a symbolic animal and divinities from the Hindu pantheon. Awakening the consciousness of a cakra (by grace, chance or through various yogic practices) brings about a new level of energy and awareness in the practitioner. Several schools of thought among the scores of new alternative practices use the cakra paradigm to help their patients/clients evolve at all levels. For many, this paradigm is the result of inner transformation in Yoga practices, because they can lead to absolute liberation (Moksha).

Each cakra has a higher energy vibration that the one before it, in ascending order. This does not mean that the lower cakras have less value, however. Grading the value of this psycho-energetic system's concepts is not the point. In fact, contrary to certain claims, awakening the cakras is not a linear experience. Practitioners can just as well open their cardiac lotus before their perineal lotus. The experience appears to vary from individual to individual.

There is a kind of radiation or emanation around these cakras. This emanation, visible to the trained eye, is expressed in numbers of corresponding lotus flower petals, which demonstrate the vibrational aspect of the cakras. For each petal, a precise vibration is produced by a sound corresponding to a letter in the Sanskrit alphabet. The first cakra, Mûlâdhâra, has four petals and therefore the lowest vibration. The last cakra, Sahasrâra, has one thousand and eight (1 008) petals and therefore has the strongest vibrational expression of the cakras. These vortices emanate both toward the front and back of the subtle body while the seventh emanates upward.

Description of the Seven Cakras

The seven cakras, in ascending order in the central axis of the vital sheath, are:

– **Mûlâdhâra-cakra**: located at the base of the spinal column, exactly at the central tendon of the perineum, directly in front of the anus and behind the scrotum in men, in front of the anus and behind the vaginal opening in women. This cakra is traditionally associated with the earth element and sense of smell. Its action organs are the feet. Its mantra is *lam*. Its animal is the elephant, symbol of strength, and its divinities are Brahma, the creator, and *Dâkinî*, an aspect of Shakti, the Universal Energy. Mûlâdhâra supports the other cakras above and is the receptacle for divine energy, *Kundalinî Shakti*, in the human structure. Mûlâdhâra is the starting point for the main nâdî, *Sushumnâ*, which rises uninterrupted along the central axis, piercing each cakra on its way to the seventh, Sahasrâra. It is also the starting point of the two nâdîs, Idâ and Pingalâ, which flank Sushumnâ on either side of the spinal column up to the sixth cakra (see the section on the three main nâdîs a little further). Its vibrational emanation is the most simple, forming four rays or petals, symbolizing the four directions of space.

– **Svâdhishthâna-cakra:** located at the root of the genital organs. It is associated with the water element and sense of taste. Its action organs are the hands. Its mantra is *vam*. Its animal is an aquatic chimera and its divinities are *Vishnu* (the Preserver) and *Râkinî*. This cakra is linked to desire (such as sex drive) and its vibrational emanation has six rays or petals, symbolizing the affliction emotions: *lust, anger, greed, illusion, pride, and envy.*

– **Manipura-cakra:** some texts place this cakra at the umbilicus, others a little higher. It is often related to the solar plexus, which anatomically, is a mass of tissues belonging to the autonomic nervous system, located around the abdominal aorta, directly above the superior mesenteric artery. This cakra is associated with the fire element and sense of sight. Its action organs are the digestive organs. Its mantra is *ram*. Its animal is the ram, symbol of fiery energy, and its divinities are *Rudra* and *Lâkinî*. Its vibrational emanation has ten petals.

– **Anâhata-cakra:** located at heart level, this cakra has twelve petals and is linked to the virtues of universal love, compassion, calmness, gentleness and kindness. Waking this centre to individual consciousness makes it easier to harmonize the other centres and avoid the potential pitfalls associated with opening the upper or lower centres. It is associated with the air element and sense of touch. Its action organs are the genitals. Its animal is the black antelope, symbol of agility and speed. Its mantra is *yam* and its awakening allows the practitioner to perceive the transcendental sound inherent to the Essence (*nâda*). Its divinities are *Îsha* and *Kâkinî*.

– **Vishuddhi-cakra:** located at the throat. This centre is associated with the element of ether and sense of hearing. Its action organs are the mouth and skin. Its animal is the white elephant, symbol of pure power. Its mantra is *ham* and its divinities are *Ardhanarîshvara* and *Shâkinî*. Its emanation has sixteen petals, one for each vowel in the Sanskrit alphabet.

- **Âjnâ-cakra:** located in the centre of the brain around the pineal gland, at the point between the two eyebrows. It is often called the "third eye". Cerebral physiology has identified light-sensitive nerve cells in and around the pineal gland, even though no light penetrates this deep in the brain. The answer may lie in the phenomenon of inner illumination claimed by mystics while in a state of ecstasy. When awakened, this centre becomes an organ of extrasensory perception, telepathy and clairvoyance. Its sense is the thinking mind and individuation (manas and Ahamkâra). Its mantra is *om*. Its divinities are *Parama-Shiva* and *Hâkinî*. Its emanation has two petals, symbolizing the polarity of the universe. *Âjnâ* is the final end point of the two nâdîs, *Idâ* and *Pingalâ*, while the central Sushumnâ nâdî continues to Sahasrâra. No animal, but rather a phallus (*linga*) in an inverted triangle (*yoni*), symbolizing the union between the male and female principles or the opposite polarities of the manifested universe.

- **Sahasrâra-cakra** or the 1 008 petal lotus, is located above the skull. It was named for its intense radiance when awakened. It is the reason why, in every culture's mystical and religious iconography, saints and sages are depicted with a bright halo crowning their head. This psycho-energetic centre is a light structure whose emanations rise up to infinity. Its action organ is the brain. Its sense is the higher mind, capable of discernment and discrimination (Buddhi). Its divinities are the coming together of Shiva and Shakti. This ultimate reunion is consummated when Kundalinî Shakti has risen to this final level. Kundalini can rise to that level because it flows in Sushumnâ nâdî, the only nâdî to reach this lotus. The result is *enlightenment.*

All of these details about cakras and nâdîs were developed as explanatory roadmaps for a subtle territory to be explored. By applying the various practical yogic techniques, yogis and yoginîs can, according to preference, come to fully know these psycho-energetic centres. Iconography, gods and goddesses, mythology, colours and symbols are all tools designed to help the practitioner more serenely integrate the vital body experience, an experience that is so fantastically extraordinary that it would be difficult to interpret out of context or give rise to understanding bordering on schizophrenic hallucination.

Cakras and the Physical Sheath

The six cakras and Sahasrâra are also related to different anatomi-cal structures in the physical body, as if these areas or organs were subtler gateways or emanations providing access to this mysterious, immaterial vital sheath. Several authors have found a connection between the autonomic nervous plexi, endocrine glands and these cakras. Despite a few contradictions, there is undeniably a cor-relation between the physical sheath and the being's other subtle dimensions.

My personal study, corroborated by other researchers, has led me to believe that the vital sheath is linked to the physical sheath by its strong natural affinity with the bioelectromagnetic fields. In fact, this problem, if it can be seen as such, is completely resolved when we look at it from another angle. As suggested earlier, a human's morphogenesis, initiated at the embryonic stage (during fertilization) is only the inescapable development of the potential of two prânic thought forms meeting. Each thought form is transported by its initial carrier (spermatozoid or ovum). The merging of these prânas triggers an evolutionary process, the individual's ontogenesis.

Therefore, the subtle body would precede the arrival of the physical body. The subtle body, the prânic thought form, is the fluid framework on which the structure's morphogenesis rests. Explained more simply, the initial aphorism *Structure Governs Function* becomes *Prâna Generates Mind, Mind Engenders Structure, Structure Governs Function.*

The Three Main Nâdîs

There are three major nâdîs along the central axis of the vital sheath: *Idâ, Pingalâ and Sushumnâ.*

Sushumnâ is probably the most important one. The axial pathway at the core of Prâna-Maya Kosha, it stems from the centre of the Mûlâdhâra cakra and travels through all the centres up to the one above the skull called Sahasrâra, the lotus of 1 008 petals!

It is said that Sushumnâ pierces the cakras. When awakened through Yoga, dormant Kundalinî-Shakti, which is coiled like a snake in the Mûlâdhâra cakra (also called *the Serpent Power*), rises and penetrates in Sushumnâ. In passing through the cakras, this fundamental prânic energy wakens its latent attributes, awareness occurs and the individual experiences a different level of self. As mentioned earlier, this ascension of Kundalinî is not linear. One cakra can be awakened before another, but each time one does, Kundalinî Shakti is at work and is the cause.

Idâ also originates at the Mûlâdhâra cakra. This energy channel winds around the central axis, beginning and ending its journey on the left side of the vital body, the female side. The Prâna that is circulating in this channel thus carries the lunar attributes: it is cold and glacial blue. In Chinese medicine, it would be called *yin*. It crosses Sushumnâ at each cakra and comes to an end in the left nostril, to the left of the sixth cakra (Âjnâ).

Pingalâ originates at the right side of the Mûlâdhâra. It is the solar aspect, male and burning with vital energy. It is often represented by a warm reddish or yellow-orange hue. In Chinese medicine, it would be *yang*. It comes to an end in the right nostril, to the right of *Âjnâ*, having crossed Sushumnâ at each of the other cakras.

In Chinese medicine, this triad of nâdîs corresponds to the *extraordinary meridians* called the *Governing Vessel* and *Conception Vessel*.

In the vital body, the primordial triad of Idâ, Pingalâ and Sushumnâ forms the central axis. The iconography most often represents these as a wide central pathway, with Idâ and Pingalâ winding around it. Other images show Idâ and Pingalâ as vertical channels flanking Sushumnâ. In reality, this triple axial channel is much finer than illustrated. Some authors say that it is no wider than a strand of hair. It should be noted that Sushumnâ can also be broken down into even more subtle channels, overlapping like cylinders in the manner of Russian nesting dolls and making up the different aspects and potentials of Kundalinî Shakti's rising.

Scope of Prâna

Since Prâna stems from Shakti, it is imbued with divine intelligence and consciousness. That is the experience claimed by all practitioners who have felt Prâna's flow and studied it through Yoga's kriyâs. It is the easiest way to describe its magnitude. Prâna is much more than a liquid, much more than a gas (some authors have tried to relate it to circulating oxygen or carbon dioxyde). When it flows with consciousness, all aspects of the person's life are magnified and potentiated.

Prâna manifests not only the vitality of the structure, but also instills meaning. The more refined its fluidity, the more knowledge and creativity will be manifested. Prânashakti thus serves as the fundamental muse, the illumination responsible for all artistic and scientific creative endeavours.

Prâna and Hatha Yoga

According to the Tantras, Hatha Yoga's primary goal is to prepare the physical sheath for optimal prânic flow. This makes it
the ultimate physical conditioning, resulting in supreme physical
fitness and health.

Most modern Hatha Yoga schools teach a series of âsana
sequences that engage the entire physical body. Regular practice
will, at minimum, put the practitioner in excellent shape. When
performing yogâsanas, each pose produces an original and unique
prânic flow. This would explain the large number of âsanas (over
64 000!) and the many schools. By incorporating other aspects of
Yoga, Hatha Yoga, a thousand-year old tantric Yoga practice, has
become a complete practice in its own right. It prepares the being
to experience the ultimate illumination (Moksha) at all levels of
manifestation.

Kriyâs of Prânic Influence
in the Physical and Vital Sheaths

Prâna is an intelligent and conscious vital fluid. It can be influ-
enced, modified or disrupted in a number of ways. Yoga teaches
some of the most efficient methods for maintaining the physical
and vital body.

Âsanas

In the first chapter, we explored the different phases of a yogâsana,
one of the favoured means of influencing prânic fluidity. Each
movement of the physical sheath can be transformed into an âsana.

Hatha Yoga is the practice of âsanas. In the West, it is the most
well known and popular form of Yoga. Most Westerners embark
on the yogic adventure through this Yoga system.

The classic characteristics of Hatha Yoga poses and the correct way to perform these do not vary greatly from school to school. Most will recognize the method of practice in this list:

- The âsana must be stable.
- The âsana must be comfortable.
- The mind is concentrated (dhârana).
- The parts of the body are always properly aligned.
- During active phases, movements are performed fluidly and with ease and only the muscles needed for these movements are engaged.
- During passive phases, only the muscles needed to maintain the pose are engaged.
- Respiration must be smooth and even at all times.
- All âsanas sequences must include poses and counter-poses.
- All âsana sequences should start and end with an integration phase, often in savâsana.

Mudrâs

Mudrâs are an aspect of tantric practices that date back to the Vedas. They are symbolic gestures that mainly involve the hands, but also include physical poses. They form kriyâs that accompany and complete some âsanas as part of a more ritualistic Yoga practice.

The energetic action of the mudrâs is owed to *vyânavâyu* (see later, in the section on vâyus). These mudrâs, like bandhas, affect prânic fluidity. Most mudrâs are energy loops that seal a flow (mudrâ means *seal* or *sealed*) or put various prânic streams into contact for the purposes of potentiation. They also have undeniable therapeutic benefits. *Anjali-mudrâ, Jnâna-mudrâ, Dhyâna-mudrâ* and *Khecârî-mudrâ* are among the best known mudrâs.

Bandhas

In the first chapter, I mentioned that muscle cocontraction around certain areas of the central axis creates a *Bandha*. Let us delve a little deeper.

There are 3 main bandhas in Hatha Yoga: *Mûlabandha, Uddîyânabandha* and *Jalandharabandha*. When applied simultaneously, they form *Mahâbandha*.

1) Mûlabandha is a perineal contraction. It specifically involves the central tendon of the perineum, located in front of the anus, behind the scrotum in men and in front of the anus up the posterior vaginal wall to the cervix of uterus in women. This contraction is different from the one involving the anus, which is another Yoga mudrâ: *Ashvinî-mudrâ.*

Performing this bandha on a regular basis results in increased energy control (because it activates the Serpent Power, Kundalinî) as well as numerous therapeutic benefits for the physical body (the prostate, for example).

Mûlabandha can be practiced with the âsana and maintained throughout a Hatha Yoga session, potentiating the practice further. Respiration should remain completely smooth and even.

2) Uddîyânabandha is a contraction of the abdominal muscles of the anterior abdominal wall, pulling the stomach back and up, under the diaphragm. It can be practiced quite intensely, forming a cavern under the ribs or more gently, only toning the anterior abdominal wall.

This bandha should not hinder the evenness of the respiration, which will tend to occur at the thoracic level. It can be practiced, like the previous bandha, during the entire âsana session. It is said that Uddîyânabandha helps maintain Mûlabandha by creating an inverted pressure, a kind of suction within the abdominal cavity, which pulls the pelvic floor up from the inside.

Uddîyânabandha is also a preparation for *Nauli* kriyâ, one of Hatha Yoga's six purification practices proposed by certain authors, grouped under the term *Shatkarmas* (see Shatkarmas section).

3) Jalandharabandha is applied by a contraction of the flexor muscles of the upper anterior cervical region. This movement brings the chin in the direction, and often in contact with, the jugular notch and manubrium sterni. Respiration should remain smooth and even in the abdomen and torso.

Applying a bandha focuses the vital energy (Prâna) to help it better circulate in the physical structure and in the 72 000 nâdîs of the vital sheath. When the bandha is activated, prâna concentrates around the central axis and potentiates the body.

The Tantras and yogic texts mention that practicing bandhas encourages purification or liberation of *granthis* (various bioenergetic blockages) that can restrict the free flow of Prâna.

Mûlabandha corresponds with the Mûlâdhâra cakra, Uddîyânabandha with the Svâdhishthâna and Manipura cakras, Jalandharabandha with the Anâhata and Vishuddhi cakras.

When Mahâbandha is applied, prâna and apâna vayus (see a little further) are invited to merge. By energetically sealing the ends of the central axis, the bandhas encourage Prâna to rise as Kundalinî within Sushumnâ.

Drishti

Drishti is a directed focus of the gaze. It is sometimes practiced by itself, as a concentration exercise, but is mainly used in the practice of âsanas to bring attention to a specific part or area of the body. It prevents mental dispersal by focusing the gaze. Prâna is directed through this concentration.

All Other Human Activity

It is possible for yogis and yoginîs to turn all activities into a Yoga practice, thus making daily life a constant celebration. From the humblest to the most elevated activities, all are pretexts for transcendence and communion with the divine. This Yoga is an integral Yoga, as Sri Aurobindo stated in the last century, and as did many other great Yoga sages before him.

Prânâyâmas

If there is a way to influence Prâna, it is through prânâyâmas. Prânâyâma Yoga is the scientific art of mastering the breath. It is an art because it can be modified into infinite variations. It is a science because it can be observed and reproduced. Research has been conducted on breathing practices for millennia, with the results passed down in the various Yoga and medical texts.

The word prânâyâma is often plural, because there are several types. Each has an explicit effect on every level of a person's being. There are prânâyâmas for health and physical fitness, to awaken Kundalinî or develop new abilities. There are also prânâyâmas to support or complete certain practices in other forms of Yoga. (In the second part of this book, which presents Yoga therapy, I will explain the therapeutic benefits of the prânâyâmas).

Prânâyâma Yoga is the Yoga of Breath. Some yogis only practice this type of Yoga.

In a rather elaborate morning sâdhana, yogis and yoginîs could begin with a series of various prânâyâmas (approximately 10 to 45 minutes), followed by a meditation session (20 to 60 minutes) and finally, a Hatha Yoga session (30 to 90 minutes).

This kriyâs sequence, in which the prânâyâmas precede dhyâna, is often said to be ideal, not only because the prânâyâmas help control prâna, but also calm the mind, making it more conducive to meditative transcendence.

Types of Prâna: Vâyus

According to the texts, there are at least ten main types of Prâna in the body-mind. Of these, five are considered more important and are viewed as five different properties or abilities of Prâna flowing in the three bodies. These prânas are perceived as breaths (*vâyus* or *prânavâyus*) that are responsible for certain physiological and mental functions.

Prânavâyu

This prâna is found in and circulates within the head. It brings outside Prâna in, flowing down and inward. Prânavâyu relates to the upper cerebral functions. Lower down, it descends into the throat, bronchial tree, lungs and diaphragm. It is therefore responsible for the inspiratory movement or inhalation (Puraka).

Prâna is the entering movement, afferent. It blends cosmic external Prâna with internal prâna. It is the movement of the mind, thoughts, feelings, emotions, sensations and perceptions.

Apânavâyu

This prâna is mainly present in the pelvic cavity, in the colon, rectum and urinary tract. In women, it is present in the uterus, vagina and ovaries. In men, it can be found in the testicles, seminal vesicles, prostate and urethra. It regulates the kidneys and bladder, as well as menstruation, defecation and flatulence. It flows down and outward.

It is responsible for the forces at work during pregnancy and uterine contractions upon giving birth, as well as the ejaculatory movement and the movement of the spermatozoids.

During coitus, the man and woman's apâna merge and cause the spermatozoid and ovum to meet. The apâna is therefore responsible for conception.

All sexual dysfunctions are caused by disruptions in apâna's flow.

Apâna is Prâna's exiting movement. In this sense, it governs *Recaka*, exhalation.

Udânavâyu

This is the ascending vital energy. Located in the diaphragm, it moves up through the lungs and bronchi, trachea and throat. It enables the movement of the diaphragm and intercostal muscles and favours expiration. As such, it controls the voice at the larynx as well as eructation.

While prânavâyu brings air and oxygen into the lungs, udânavâyu expels carbon dioxyde through its ascending movement. It therefore favours good oxygenation. It also penetrates the brain to stimulate memory.

During exhalation or Recaka, apâna expels the Prâna from the body, but udâna carries it up, from the diaphragm to the head, by way of the larynx.

Samânavâyu

This is the vital breath in the small intestine, around the umbilicus. Samâna is therefore responsible for the motility of the duodenum, jejunum and ileum. It is the force behind the secretion of digestive juices. It is therefore directly related to the "digestive fire". It is also responsible for hepatic and pancreatic secretions as well as peristalsis of the small intestine.

What is more, samâna is the force behind hunger pangs. It governs appetite, digestion, intestinal absorption and assimilation.

Vyânavâyu

This type of Prâna mainly maintains cardiovascular activity, blood circulation, nutrition and oxygenation of the tissues and organs. It is also responsible for venous and lymphatic return and finally, all circulating fluids. It flows through all of the connective tissue fibres of the physical sheath. It is also the vital force behind the neuromusculoskeletal reflexes. In short, it is a vital force found throughout the connective tissue framework of the physical body, spreading to every level.

When practicing prânâyâmas, yogis and yoginîs can join prânavâyu and apânavâyu with Mûlabandha in the mûlâdhâra cakra. Bringing these two opposing vital forces together is said to help awaken dormant Kundalinî Shakti in this area and make it ripple up in Sushumnâ nâdî. Practitioners would do well to learn the details of these techniques and practice them under the supervision of an experienced yogi or yoginî.

The five other types of Prâna are considered secondary and therefore less important:

- *Nâga* (serpent), responsible for vomiting and eructation.
- *Kûrma* (turtle), responsible for the movement of the eyelids.
- *Kri-kâra*, responsible for hunger, thirst, the urge to sneeze.
- *Deva-datta*, responsible for yawning or sleep.
- *Dhanam-jaya*, responsible for the disintegration or decomposition process or necrosis.

These ten prânas form the Prâna that maintains the body-mind. Most commentators agree that prânavâyu and apânavâyu are the two most important, given their continuous activity tied to respiration. When the mind is agitated, prâna and apâna are agitated and vice versa. The importance of prânâyâmas is clear: controlling prâna and apâna will calm the mind (see the later section on the mental body, Mano-Maya Kosha).

It is therefore vital for experienced yogis and yoginîs to learn and practice Prânâyâma Yoga in order to bring about this sought-after mental pacification and add a wonderful means for self-realization to their toolbox.

Vasi Yoga

Vasi Yoga is the ultimate Dravidian Siddhanta Yoga (from the southern part of the Indian subcontinent). It is an esoteric form of Prânâyâma Yoga focused on the release of Kundalinî, practiced by siddhas (practitioners of Shaivite tantric Yoga in India). Today, propagators of Dravidian Kriyâ Yoga mainly teach this Yoga. According to its practitioners, it is the quickest means of bringing about an awakening. It accelerates spiritual evolution while helping the physical body harmoniously hold the liberating Serpent. Vasi Yoga is a type of sacral spinal respiration that calls on the three bodies. It can only be learned through *dîkshâ*, personal initiation between master and disciple. Over three thousand years ago, *Tirumûlar*, one of the most illustrious Dravidian siddhas, praised it in his treatise, *Tirumandiram*.

Introversion of Senses

Hatha Yoga and Prânâyâma Yoga lead practitioners to gradually narrow their sensory perception, which is a paradox, since these Yogas make the body and its senses even more receptive. However, in yogic practice, the senses are slowly brought inside. In this way, distractions diminish while *attention* arises.

According to Pantajali, this gradual, voluntary sensory introversion, which he calls *Pratyâhâra*, is one of the eight milestones of self-realization and often the prelude to *Dhâranâ*, the practice of concentration.

Prâna and Emotions

The flow of intrinsically intelligent Prâna nourishes mental activity and helps mindfulness. I mentioned earlier that Prâna is the source of thought. Let us explore that idea.

Attention or mindfulness is a type of focus, a concentration of the mental activity. It leads to very sharp perception. Perception is linked to the senses (indriyas) and leads to sensation. Sensation leads to feelings. Feelings are a somewhat complex and persistent inner state, emotional in nature, pertaining to our interpretation of a thing or experience. This chain of mental events triggered by Prâna produces a thought, an inner image with which the person resonates. The feeling becomes an emotion.

Prâna-maya kosha becomes thus the priviledged sheath for the emergence of emotions. That is the natural order.

The Inferior Mental Sheath - Mano-Maya Kosha

In his introduction to the Yoga Sûtras, Patanjali defines Yoga as the *cessation of fluctuations of consciousness* (I-2).

Consciousness

Human beings are endowed with consciousness.

This notion of consciousness has been a fertile topic in philosophy through the ages and has been actively studied in the cognitive sciences in the latter half of the 20th century. Without claiming to have the final say on this complex concept, I will present a few ideas from the yogic tradition.

Tantric Anthropomorphism and Consciousness

The notion of consciousness first exists as a basic principle, the essence of all there is. It encompasses the philosophical idea of a primordial source from which everything emanates, but it can also be found in religion, with the notion of the All-Powerful God, Eternal Creator of the Universe. The concept is anthropomorphized in the different Christian traditions: God, the Father, with his long, white beard, imposing power and wisdom, the Great Creator. In the Muslim tradition, it is Allah. In the yogic tradition, this concept is anthropomorphized again. In Tantrism: the unchanging Shiva. In Hinduism: Brahman.

Let us randomly stroll down the tantric path.

When Shiva manifests, it becomes Shakti, the all-powerful Mother energy and Creator. Tantrism is a celebration of the primordial energy of Shiva. Shiva is Pure Consciousness, *Cit* in Sanskrit. His primal movement is Shakti, the initial Pure Vibration.

Shakti is the force behind creation, Nature, Prakriti. In the living world, Shakti is Prâna. *Prânashakti* therefore creates human beings.

Human beings first vibrate in Ânanda-Maya Kosha, the human home of the soul or Jîva, which is Shiva's image. It is the most ethereal body, radiating its energy both to the infinity of Nature and at the heart of the being. Its vibrational frequency slows somewhat, which is when Vijnâna-Maya Kosha and Mano-Maya Kosha appear, the mental being, the mind, whose vibration is essentially limited to the space around the physical body. When the vibration continues its contraction, the result is the crude structure of Anna-Maya Kosha, the physical body.

Prâna-Maya Kosha interpenetrates and connects the physical and mental bodies to Ânanda-Maya Kosha.

Cit (consciousness) and Citta (individual consciousness)

When consciousness (Cit in Sanskrit) vibrates and begins to move in Jîva, the mind is created, with Prâna leading. Prânashakti's movement creates and maintains life in this body-mind.

The human mind (or mental sheath) cannot be separated from its physical component. The mind exists because it is linked to an incarnated body and that physical body cannot exist without its mental component. It is this idea that is conveyed by the concept of "body-mind". The body-mind is an indivisible unit. Prâna-Maya Kosha connects Anna-Maya Kosha and the two sheaths of the mental body. Without the vital principle that binds them, the physical body is only a worthless heap of inert waste and the mind is a lost immaterial entity.

The mind is the mental aspect of an individual. It also shows a consciousness, the individual consciousness, or *Citta* in Sanskrit. Commentators have broken down this individual consciousness in several ways. I will mainly concentrate on the tantric perception of the mental body and occasionally attempt to make some connections to Western philosophical concepts.

The yogic tradition and tantras have divided the consciousness of the human mind into three parts: Manas, Mano-Maya Kosha or the lower mind, Ahamkâra or ego, and Buddhi or the higher mind. Ahamkâra and Buddhi form Vijnâna-Maya Kosha.

Functions of Manas in Mano-Maya Kosha

Manas forms the least evolved part of the mind, also called the lower mind. Its key role is sensory interpretation. The ten powers of knowledge (the five classic senses and five agents of cognitive action) are supported by manas, which acts as the being's primary experience integrator.

In the physical body, the sensory axis of the nervous system, the limbic brain, the reptilian brain and the thoracic spinal cord drive the autonomic nervous system. The physiological functions that these structures control resonate in synchronicity with manas. Autonomic nervous activities are therefore determined by the joining of Prâna-Maya Kosha and Mano-Maya Kosha in Anna-Maya Kosha. This resonance can spread to all the fibres of the physical body due to its immense connective receptivity, as we saw in the first chapter.

Citta's lower mental activity takes place when the latter is stimulated by a sensory nerve impulse originating from the sense receptors. This stimulation not only creates the autonomic physiological process, but also produces thought. A thought is an image or concept. Thoughts also have an echo effect on the autonomic physiological functions. Mano-Maya Kosha only records and interprets this stimulation and produces thought. That is its main function. The texts claim that manas is an eleventh sense.

Manas thus causes images, sensations and concepts to arise in the individual's consciousness. It does not judge or evaluate; it simply reports. In this way, it stores information, it memorizes. This memory forms the basis of the person's learning and shapes physiological development because the being reacts to this information.

The individual mainly responds to this constant sensorial and sensual bombardment via a set of motor reflexes and physical actions based in the unconscious. The person's functional knowledge and evolution therefore stem from experience, the past, and form the subconscious root of the person's beliefs and behaviours, determining how he/she will react in the present and colouring decisions made in the future.

Samskâras and Vâsanâs

Every experience leaves an impression in manas as well as in Buddhi, the intellect, which we will explore soon. Mental traces are subliminal activators that form the basis of the yogic concepts of *karma*. They are called *samskâras* in the yogic tradition.

All sensations, thoughts, feelings, volition or any other type of conscious and unconscious creation form a samskâra. All samskâras of the same type are recorded and form *vâsanâs*, tendencies concerning this thought or action, which are buried in the mind and characterize the embodied mental being.

The set of *vâsanâs* make up the uniqueness of the human mind, its ego personality. It is only by dissolving the vâsanâs or the set of samskâras that we can stop the ceaseless karmic wheel of successive incarnations, which is often the ultimate goal of yogic practice: illumination or moksha. This is the Yoga Patanjali speaks of in the introduction to his Sûtras. Tantric yogis believe it can be achieved in this lifetime and in this physical body.

The Intellect (Buddhi) - Vijnâna-Maya Kosha

While Mano-Maya Kosha records, coordinates and files sensory information and memorizes experiences, Vijnâna-Maya Kosha forms Buddhi, the higher mind. The two terms can be used interchangeably.

Intellectual function dominates Buddhi. Its power is the light of intellect. It is the centre for reasoning, evaluation, judgement and discrimination. In short: intelligence.

Buddhi illuminates the mental process. In its highest expression, it is mental alertness.

Ahamkâra

By differentiating concepts, Vijnâna-Maya Kosha appropriates each experience and makes it personal, unique and separate from another person's experience. From this springs the "I", the mental concept of the ego. Egocentrism is the often unconscious tendency to make everything about oneself. This automatic mental faculty is the ego, or *Ahamkâra*.

In other words, Ahamkâra introduces the separateness of subject and object to the person's consciousness. This causes the emergence of the roots of experience: the ten senses (indriyas), the five potential subtle energies (*tanmâtras*, sources of sensory stimulation) and the lower mind (Mano-Maya Kosha).

Personality and Beliefs

The union between Mano-Maya Kosha, Ahamkâra and Buddhi's higher mental functions create a human being's personality. It is characterized by the affirmation of the person's beliefs, which themselves are based on memories.

A person's beliefs are also anchored in the unconscious mind. Beliefs stem from culture (environment) and impressions left by an experience.

An individual's personality (including the emotions) is the main trait of the merging of the individual consciousness, Citta, and the physical sheath.

Unconscious

The activity of the mental body, Mano-Maya Kosha and Vijnâna-Maya Kosha, essentially occur unconsciously. Some call this the *subconscious level* because this unconscious part is accessible and can be transformed through yogic practice. It is important to understand that the greater part of our consciousness (Citta) is in fact unconscious. Only the Buddhi portion occasionally expresses itself consciously, essentially in acts of volition. We can optimize and make our *consciousness more conscious.*

If we understand that our mental sheath regulates our physical sheath's activities, we must conclude that the vast majority of our behaviours and beliefs are rooted in the subconscious. We can therefore see that our normal consciousness when awake is in large part made up of automatic responses.

Manas records and stores data throughout a person's lifetime, continuously building a mental construct of reality as perceived by the senses. Ahamkâra is the part of Buddhi that appropriates the sensory and emotional experiences and makes them personal.

Vijnâna-Maya Kosha refines this construct, these experiences, by differentiating between those that produce the most pleasure and joy. It judges, evaluates and discriminates, directing thought—consciously, but most often unconsciously—toward physical action or intuition and introspection.

All of this chiefly unconscious activity shapes the being's mental construct, which is made unique by the person's karmic traits (vâsanâs) and personality.

Again, the majority of the functions described above take place unconsciously. Certain functions of the Buddhi (awareness, decision-making), rise to the consciousness, but always remain firmly rooted in the unconscious. This is why our reactions, motivations and even beliefs are not always easily understood by Vijnâna-Maya Kosha's reasoning and logic. Yogic practice gives access to this subconscious sphere and offers endless possibilities for acting on and controlling our motivations and highest aspirations.

Emotions

We saw earlier that, in its natural order on the mind, prânic function (Prâna-Maya Kosha) produces emotions. It expressed itself both in the mental structure and the physical structure, in the body-mind.

Emotions are thus produced at the intersection of four sheaths (physical, vital, mental and intellectual body).

The emotional body some psychotherapists speak of is the product of agitation and resonance between the prâna of the vital body at the cardiac, abdominal and cerebral levels with the three other sheaths (Anna-Maya-Kosha, Mano-Maya-Kosha and Vijnâna-Maya-Kosha). In other words, emotions arise through the quadruple conjunction of the sensory experience (physical) with the memory (mental), intellect and the power of vital energy.

The emotional being also has roots in the subconscious.

Conscious Buddhi

In its highest conscious expression, Buddhi is inclined to transcend.

Intelligent and aware beings will inevitably be drawn to follow a path that leads to ever-greater realization. This will come as an irresistible urge in which Buddhi paves the way for ascension, an access through meditation, prayer or any other inspiring activity, and thus a higher inner vibration. That is when consciousness penetrates Ânanda-Maya Kosha and meets *Jîva*. Shiva and the being then become one.

Kriyâs That Influence the Mental Sheaths

Yoga offers a multitude of somewhat ritualized psychophysical tools, or kriyâs, to improve prânic flow in the mental sheath, Mano-Maya Kosha, and the intellectual sheath, Vijnâna-Maya Kosha.

These are some of the most important kriyâs:

Prânâyâmas

Practicing prânâyâmas gives the physical sheath superior lung capacity and oxygenation. But it is also one of the most effective tools for quieting the mind.

The yogic tradition developed the scientific art of mastering breath over two thousand years ago. The resulting Yoga practice is Prânâyâma Yoga. The benefits recorded during cardiac coherence are exactly the same as during a gentle practice of Sukha prânâyâma; which is only one of the most accessible known prânâyâmas.

Prânâyâmas induce peace of mind and prepare practitioners to explore the depths of being through meditation.

Intention

Intentionality is an important function of will. It is the prelude to manifestation. When the mind is continuously focused on an objective, a goal, the intention modulates and guides prânic flow. Concentration or Dhâranâ drives intention. In order to create and manifest, it must be pure. This incisive state is obtained when the mind is clear (no confusion). That is when the subconscious and conscious parts of the mind are perfectly aligned, manifestation thus become automatic.

Dhâranâ

Dhâranâ is the practice of concentration, keeping the mind immobile, in a state of non-action. It is often preceded physically, by sensory introversion.

There are several types of such kriyâs in the yogic practices. They all aim to strengthen the ability to focus the inquisitive mind (internal attention) on a single object, whether material (with form, *rûpa*) or immaterial (formless, such as a concept, *arûpa*).

Dhâranâ is a wonderful yogic practice that goes beyond the simple act of intellectual concentration. As a yogic practice, it concerns the entire being. This means that the physical body can be as focused as the mind, not a muscle engaged, not a single sensation interpreted. Dhâranâ is pure and total.

Dhâranâ is often the prelude to dhyâna.

Dhyâna

Dhyâna is the *scientific art of mastering the mind*. It is usually translated as *meditation*.

All yogic and philosophical traditions have developed this tool for transcendence.

True meditation occurs when the object of concentration completely fills the consciousness; nothing else exists. The body is unaware of the rest of the inner or outer world, only the object of meditation. The person meditating is the explorer and observer of his/her unique object of inquisition. In this state, Prâna circulates even more efficiently.

If the object of the dhyâna is an indriya, the meditator will gain complete knowledge or siddhi (perfect understanding and mastery) of this indriya.

Dhyâna is often the prelude to samâdhi.

Dhyâna Yoga is another complete form of Yoga.

Samâdhi

Samâdhi is ecstasy, or more precisely enstasy, since it is an ultimate state of inner fusion. It is often activated or preceded by dhyâna and characterized by this state, in which all fluctuations and disturbances of ordinary consciousness evaporate through meditative absorption.

Samâdhi is the outcome of Yoga, its result. Some schools even equate Yoga and samâdhi, the two expressions conveying a union or fusion between the subject and object of experimentation.

Samâdhi can be the linear result of progressing from dhâranâ to dhyâna to samâdhi or it can be a spontaneous, unplanned event. For yogis, it is the ultimate achievement.

While it is not an easy practice, because samâdhi is not often taught or shown, it remains (at least in its initial stages), a practice. And, as such, it is another kriyâ promoting greater prânic fluidity.

In classical Yoga, there are several stages of samâdhi. The initial stages retain certain traces of thought consciousness. The later stages eventually lead to a release from the cycle of life and death, to absolute realization, Kaivalya.

Communication with Kârana Sharîra (the causal body) is easiest through samâdhi.

When Prâna is stabilized through âsana, prânâyâma and dhyâna, it again becomes pure self-awareness, pure serene potential. It returns to the source, Ânanda-Maya Kosha, from which it sprang.

CHAPTER III

KÂRANA SHARÎRA
THE CAUSAL BODY

The Spiritual Sheath or Sheath of Bliss
Ânanda-Maya Kosha

Introduction

This third chapter completes the first part of the book. It only contains one section since Kârana Sharîra, the causal body, only has one sheath. I now present Ânanda-Maya Kosha, home of the soul and fundamental source of the being.

As in the previous chapters, I will examine the main Yoga kriyâs liable to benefit this body.

Ânanda-Maya Kosha and Jîva

Kârana Sharîra, the causal body, is the inner temple. Its sheath is Ânanda-Maya Kosha. The heart of this sheath, its jewel, is the soul, *Jîva*.

This is where all logical reference ends. We are now entering an immeasurable dimension of the being. Researchers, seeking demonstrations and tangible, quantifiable, irrefutable proofs, will have to be satisfied with highly subjective reports that vary by source. There is one constant, however. The texts mention that Ânanda-Maya Kosha appears to contain a spark of what comes closest to the indefinable source: Jîva, the human soul.

In Tantra, pure consciousness is considered beyond expression and identification: Parama-Shiva, the Ultimate Reality, Brahman, God or Allah. Its subjective, unchanging aspect is Purusha, the primordial soul. We experience its objective, observable manifestation when it begins to move and Shakti appears. Shakti is Prakriti, the Universe, Matter, and Nature. Its movement is Prânashakti or Shakti manifesting. This is how the being emerges. The flame of the divine creative essence and Shiva's image form in each, the most ethereal part of the being, the soul, Jîva.

Ânanda-Maya Kosha is also known as the *spiritual sheath* or *sheath of bliss*. Its expression is a manifestation of joy, of gratuitous beatitude. Ânanda-Maya Kosha touches the essential vacuity of the being, Shiva Supreme. When Jîva is activated through intention, the various cruder bodies appear.

As soon as this consciousness begins moving, it is called Prânashakti and becomes prânic flow. Its vibration creates and infuses the body. Prânashakti is prânavâyu, and attention arises. That is how Prânashakti creates the mind; Prâna creates Manas and then Mano-Maya Kosha and Vijnâna-Maya Kosha.

When, through yogic practice, its movement stabilizes, Ânanda-Maya Kosha becomes one with Shiva again. When its movement resumes, it expresses (through Prânashakti) the subtle form of the being, Sûkshma Sharîra, made up of the three sheaths: vital, mental and intellectual, then its physical form, Sthula Sharîra, the crude physical sheath.

Ânanda-Maya Kosha and its *Jîva* are therefore the source of the being's manifestation.

Kriyâs with Prânic Influence on Ânanda-Maya Kosha

This list is not exhaustive. I only mention some of the most common Yoga practices and those that inspire me the most.

Mantras

Reciting mantras is a Yoga practice in itself. *Mantras* are sacred sounds, words or phrases, often in Sanskrit, that, when uttered, carry a vibration that raises consciousness beyond the usual states of wakefulness. If the mantra is only one sound, like "om", it is called a *bija mantra* or seed syllable.

One of the most ethereal forms of Mantra Yoga is the Yoga of the primordial sound, *Nâda Yoga*, in which the practitioner internalizes the sound vibration in order to transcend everyday consciousness.

There are bija mantras for each divinity in the Hindu pantheon as well as for the development of every possible physical and mental attribute.

This is traditionally a silent practice, often using a string of grains or minerals (a *mâlâ*) reminiscent of rosary beads used in the Christian and Muslim traditions.

Mantra Yoga has an elaborate ritualistic aspect, usually supported by devotion. It can be combined with music and become an important part of devotional chanting in Bhakti Yoga.

Bhakti

Bhakti is often translated as *devotion*. This is another practice common to many religious and mystical cultures. Its aim is a type of transcendence or ecstasy, in which the love of the divine, the mystical union (*mystical wedding* or *unio mystica*) acts as the transporting element. In this state, the individual consciousness fuses with the Supreme Consciousness expressed here by the practitioner's divinity of choice.

Bhakti Yoga is the Yoga of devotion and surrender to the divine. Some authors claim that it is the quickest path to realization. At this level, Prânic fluidity is absolute.

The most common kriyâs here are meditation, prayer, devotional chanting and samâdhi.

Joy, Compassion, Love and Equanimity

In Buddhism, the first of the four noble truths about existence is *Duhkha*: suffering is.

For Buddhists, following the path to liberation is the desire to see all beings freed from suffering. Universal, unconditional love (a variation of the Bhakti theme) causes suffering to be transmuted into a state of peace. This Buddhist concept is expressed by one of the four spiritual faculties of love, essential to relieving humanity's suffering. These are known as the *Four Immeasurables* and are: joy (*Mudîtâ*), compassion (*Karunâ*), loving-kindness (*Maitrî*) and equanimity (*Upekshâ*).

Patanjali also proposes these types of love, to obtain the same results, in his Yoga Sûtras. (1-33)

The Four Immeasurables are infectious and create, in and around the person, a perfect fluidity of the prânic flow leading to samâdhi.

Gratitude, Recognition and Ho'oponopono

Every time you thank and acknowledge the divine, you *are* the divine. It is this attitude that puts you into harmonic resonance with the Universe's Essence. Your Jîva radiates around and inside you. Being grateful is always the right path.

An original Hawaiian philosophy, called *Ho'oponopono*, was brought to my attention a few years ago and I believe that it is sufficiently inspired to be included here. This philosophy favours the resolution of inner and outer conflicts by adopting a very powerful personal attitude.

The method is infinitely yogic. It is both simple and complex.

Ho'oponopono's key concept is that you are fundamentally and completely responsible for your experience. This means that regardless of the events in your life, regardless of the state you are in, you are ultimately both its source and resolution. Nothing exists outside of you. You are the absolute source of all, positive and negative. This idea is not always easy to accept, especially in the highly emotional grips of an illness or in the face of a seemingly random traumatic event, when it is easier to fall pretty to a victim mentality.

This complete accountability has nothing to do with blame or guilt about an experience. If the experience is in your existence, you created it. Accept this simple fact and begin taking action to change it.

To solve conflicts at every level, Ho'oponopono suggests an inner attitude very similar to Buddhism's Four Immeasurables, involving the mental recitation of four fundamentally and profoundly transformative phrases. Here, I present the four phases in the framework of a confession or small elegy between you and your Essence. Apply this as your heart and instinct dictate:

— You, to your Essence:	— *I'm sorry*
Pause to breathe	
— You, to your Essence:	— *Please Forgive me*
Pause to breathe	
— Your Essence, to you:	— *I love you*
Pause to breathe	
— You, to your Essence:	— *Thank you*

The secret is not rote recitation, but putting your whole heart into it. You are sincerely sorry for causing the situation or event. You sincerely ask for forgiveness. Your Essence (God) will always hear heartfelt repentance. You will experience immediate and dazzling healing (and love) and will be filled with an urge to give humble thanks. This moment of gratitude in itself creates and heals.

Frequent repetition of this internal practice ensures its consolidation as a source of miracles.

Samâdhi

The best means of accessing your Ânanda-Maya Kosha is samâdhi. This ecstatic fusional union is, in fact, the end result of all the kriyâs mentioned in this section. Entering samâdhi often requires intense, disciplined training, the unexpected grace of a master, or a great deal of luck...

So You Think You're In Control?

Finally, it is important to remember that Prâna is intrinsically intelligent and divine by nature. Regardless of how sophisticated the yogic techniques humans develop to influence it, in all its forms, it is actually Prâna that is in charge. It shapes the way in which things occur. It is the force behind all activities and behaviours, including all mind games. It is the source of insight and intuition, all creative and artistic endeavours and the expression of love.

Do you truly believe that you are in control? Know that you are always under the spell of Prâna, that spark of Shakti.

Yoga makes you aware of this. Yoga gives you *Awareness.*

SECOND PART

INTRODUCTION TO YOGA THERAPY

CHAPTER 4

CHAPTER IV

YOGA THERAPY AND HEALTH

Introduction

In this second part of the book, entitled *Introduction to Yoga Therapy*, I suggest applying the knowledge that is provided. It will help yogis and yoginîs become agents of transformation, someone who can help heal the body-mind.

I would first like to address a few ethical notions and the need for Yoga therapists to receive adequate training.

After introducing the concept of a human being's intrinsic self-regulation, I will present a few fundamental characteristics of the physiology of illnesses (pathology), knowledge that is essential to preparing a therapeutic intervention.

This therapeutic intervention will cover every level of the being, via a protocol that explores the therapeutic approach to the three bodies.

I present a simple protocol for Kârana Sharîra, and then examine the therapeutic possibilities for Sûkshma Sharîra, especially the therapeutic benefits of the prânâyâmas.

In the next chapter, I discuss Anna-Maya Kosha, the physical sheath of the gross body, Sthula Sharîra. I introduce the concepts of pain and suffering, then list the classic and less common steps for determining the extent of a physical injury (acute or chronic), the phases of healing and the impact stress has on the physical structure. Sprains and strains are also covered. The same approach is taken with several classic chronic pathologies like arthritis, etc.

The patient must also continue the therapeutic work. For this purpose, I provide a very sure self-treatment method.

I then explore the connections between stiffness (hypomobility) and hypermobility in the physical structure and how these can be managed through Hatha Yoga.

After asserting that all injuries or illness can be improved through Yoga therapy, I include a safe method for building therapeutic asâna sequences.

In the last chapter, I explore how to prevent illness and preserve health through a regular Yoga practice. I share my strong belief about the preventive benefits of regular practice and the advantage of creating prânic fluidity in the structure first. I illustrate this aspect by demonstrating the usefulness of freeing the central axis of the body as a preliminary goal. I explore the concepts of obstacles to fluidity and physical release of tension. Finally, I insist that all Yoga therapists should become competent teachers of the Yoga methods.

Yoga Therapy and Health

Health is a natural state of well-being. On a day-to-day basis, it is expressed by abundant energy, unbridled, boundless joy and inner bliss that radiates all around. For yogis, this means being in perfect physical, vital, mental, intellectual and spiritual harmony. Yoga therapists help their clients reach this state through their knowledge of the best methods, techniques and kriyâs in their therapeutic arsenal.

Training, Skills, Ethics

All therapeutic care involves a type of helping relationship. Health care professionals attempt, through their experience, skills and all the therapeutic tools at their disposal, to meet their clients' request for help.

To play this role, professionals need to be properly educated (with supporting diplomas) and capable of demonstrating the soundness of their knowledge and training.

Most self-respecting professionals are members of a professional association that validates their competency and ensures that the professional code of ethics is respected. At the present time (2012), for Yoga therapy, the *International Association of Yoga Therapists (IAYT)* plays this role very capably in the West. However, Yoga therapy training remains the prerogative of numerous private schools around the world, which do their best, sometimes with limited means, to meet the increasingly specific criteria set during the IAYT's different plenary sessions.

Many Yoga therapists have never received professional Yoga therapy training. Many are already health professionals and experienced yogis. Some joined the IAYT and use Yoga therapeutically by incorporating it in their own practice.

That is precisely my own situation. Osteopathy is my field of medicine, with a specialization in functional anatomy. I have taught this subject here and there throughout Europe and America for over 20 years. Having personally practiced Yoga for over 30 years, I incorporated osteopathy into the yogic lifestyle. It has become, for me, another form of Yoga.

It is this expertise that allows me to humbly help future Yoga therapists and experienced yogis deepen their knowledge of the body-mind. If inspiration strikes, this knowledge may trigger the start of a new professional dimension in their lives.

For the last 15 years, I have shared some of the concepts found in this book with future Yoga teachers in teacher training classes given in schools and ashrams. As mentioned in the introduction in the first section, this book was written in response to repeated requests from my students/yogis for more knowledge about the relationship between Yoga and its therapeutic potential.

Who Consults a Yoga Therapist?

A beginner Yoga therapist's first clients are usually family and friends.

Why do, or should, people consult you? Because you have built a relationship of trust with them and they are confident in your ability to help them through Yoga therapy. You are already known for your knowledge as a Hatha Yoga instructor in a given school or style. Or you are known as an experienced yogi or yoginî and your family members, students, etc. have sufficient trust in your skill to let you help them a little further on their quest for well-being.

If your work brings positive results, you will build a reputation through word of mouth.

If you are a health care professional and already have a client base, you are adding important arrows to your therapeutic quiver. In fact, by incorporating Yoga in your therapy or medical field, you will increase its potential.

Self-Regulation and God's Pharmacy

The body-mind has the remarkable ability to self-regulate. When not pushed beyond limits by the severity of a problem, it is quite capable of managing and controlling stress and tension. Most attacks are already resolved before the pathology reaches the consciousness.

This massive and humble self-regulating force is everywhere within, from the crudest to the most subtle sheaths. In seconds, it can mobilize a horde of immune cells that release several types of biochemical responses to help contain an antigenic attack. Or it can bring about an unexpected, intuitive, freeing awareness in a mind disturbed by neurosis or suppressed emotion.

Regardless of the chosen approach, human beings have all they need to heal or maintain health, strength and power. Some of my professors call this ability *God's Pharmacy*. Every medicine's greatness lies in its ability to draw from this immeasurable resource to produce results.

God's Pharmacy is within. Its only medicine is Prâna's fluidity. Yoga therapy's aim is to re-establish this fluidity in the different sheaths. Perfect and total health requires the three bodies to be in harmony.

In their therapeutic interventions, Yoga therapists develop and apply a *treatment protocol* whose ultimate goal is maximum prânic fluidity in these three bodies.

Restoring Health

It has been said before: restoring health involves the entire being. The beauty of Yoga therapy is that it takes all three bodies into account.

The secret lies in the art of getting prânic flow in motion at every level. That is what I propose in the next several pages.

Reaching this goal takes more than words. It takes action; which is why Yoga therapists must first experiment and apply the therapeutic approach to themselves before being able to pass it to others. This is quite easily done as there is no need to be ill in order to apply the exercises that follow. Even if the Yoga therapist does not require treatment, it will still be an excellent means of growth and personal power.

Initial Awareness of the State of the Illness

Holistically, health is the being's baseline state. Health is always in the background, even when masked by illness. The art of medicine is to know how to make it resurface by stimulating the self-regulating forces. This original understanding leads us to view illness not as an imbalance, but rather, inversely, a balance. It is the best balance that the body-mind could achieve at this time. Depending on circumstances and stress, the illness is the resulting harmony of the being's current self-regulating capacity. If the attack exceeds the limits of this capacity, illness develops.

Illness Characteristics
in the Physical Sheath

When illness develops, the physical sheaths shows characteristic signs and symptoms that are now widely accepted in modern medicine. Many lay texts and specialized journal articles on this subject can be found in libraries, bookstores and on the Internet.

Inflammation

Of all the symptoms, the most significant is certainly inflammation. In fact, medical research recognizes that not only is chronic illness characterized by inflammation, but that the presence of inflammation is often one of the major precipitating causes of morbidity. In other words, inflammation is not only an important symptom of chronic syndromes, in many cases, it is the main cause of chronicity.

It is our right to question the details of this causality. That is where the explanations become vague. There are many factors involved, very often behavioral. A problem may have developed gradually, over a long period of time (sometimes decades!), making it difficult to pinpoint an event or a date on which the pathological process was triggered.

Finally, it must be noted that inflammation is the immune system's normal response. In fact, it is the first line of defence against all types of attacks affecting an individual's physiological integrity. An immediate inflammatory response will occur at the site of a small cut or bacterial attack.

Although inflammation is a normal, healthy process for acute problems, its chronic persistence can lead to a pathological state and act as the precursor to illness.

I will explain the significant role it plays in maintaining chronic illness.

Acidity = Mined Field

Inflammation can develop more easily in an acidic environment. Since people suffering from chronic conditions often have a very low pH, the two symptoms appear to go hand in hand.

A significant connection can be made between acidifying behaviours and the development of an illness. For example, prolonged consumption of refined sugar will, sooner or later, induce acidosis, in addition to disrupting blood sugar levels. Depending on the person's history and certain predispositions, he or she may develop a rheumatic condition (which can deteriorate into arthritis), hypoglycemia, diabetes and other degenerative diseases. In the worst-case scenario, a person may suffer from several of these conditions at once!

The solution to acidification is not necessarily dietary. It is not enough to simply cut out acidic foods as prevention. In fact, not all acidic foods continue to be acidifying once metabolized. What is more, some acids are vital for and support health.

The behaviours most likely to create an acidic environment are those that cause the body excessive stress: smoking, dehydration, caffeine, alcohol, refined sugar, junk food in general and processed foods in particular. Red meat is the most acidifying of the meats.

Finally, the two most important pH-regulating systems in the body are the kidneys and respiratory system. All chronic pathologies show higher-than-normal respiratory frequency at rest. This increase in ventilation is chronic hyperventilation. By eliminating an excessive amount of carbon dioxide, chronic hyperventilation produces alveolar and arterial hypocapnia, or lack of CO_2 in the tissues. This state induces oxygen retention in the red blood cells (inhibiting the Bohr effect), vasoconstriction and bronchoconstriction, thus reducing the supply of oxygenated blood to the brain and other vital organs. The resulting hypoxemia increases the anaerobic efforts of the cells and induces an accumulation of acid waste (lactic acid), which is much more difficult to eliminate than the acidity normally produced in the presence of CO_2. Finally, the lack of carbon dioxide irritates the nervous system, making it much more susceptible to chronic dysfunctions. In fact, CO_2 has a known calming effect on nervous activity.

Illness Characteristics in the Other Sheaths

When illness has set in, it confirms that the person's general condition has gone beyond the physical structure and spread throughout the entire being. The self-healing and self-regulating forces set in motion by the illness are drawing from the reserve of vital energy, Prâna.

In all dysfunctions, every dimension of the being is affected. However, in the acute phase, they are mainly confined to the physical sheath. A simple intervention in this area will correct most situations. The proposed protocol takes into consideration whether a condition is chronic or acute.

The fight waged against a chronic attack will be more serious because the being is completely affected, in every sheath. This is a daunting task. The person will quickly show signs of fatigue or even exhaustion.

By invading the entire being, chronic illness leads to ever greater vital and mental confusion. This is expressed by agitation and emotional instability and ultimately, suffering and depression (see the later section on pain and suffering).

CHAPTER V

THERAPEUTIC PROTOCOL

Therapeutic Protocol

The following points will be crucial landmarks in the effective, definite return to total health.

Yoga therapy encourages the restoration or maintenance of health by deliberately taking a global or holistic view. This involves not only treating the physical structure, but also the person's mind and soul.

The chief therapeutic constant will be to favour the emergence of optimal prânic fluidity. This result is obtained by setting Prâna-Maya Kosha in motion inside and Prâna aroud the being.

From the inside, Prâna arises like a voluntary emanation of Jîva, the personal soul, Universal Shakti's duplicate within, in Karana Sharîra. Prâna flows in Prâna-Maya Kosha and spreads both into Sthula Sharîra and Sûkshma Sharîra.

Outside, Prâna penetrates Anna-Maya Kosha through the nine gateways, the ten indriyas and the mind. It merges with the internal Prâna, invigorating the physical and mental sheaths. The nine gateways are the eyes, ears, nostrils, mouth, urethra and anus. There are kriyâs to make these gateways particularly receptive, but a correct and simple practice of Hatha Yoga âsanas will do the job.

Process for the Three Bodies

In the interest of clarity, I have divided the therapeutic process into three parts, with a suggested intervention for each of the three bodies. The process can be started on any of the bodies.

The work will cover the five sheaths already discussed in the first part of this book.

One last note: it is not absolutely vital to actively and consciously work on the three bodies for healing to occur. If Yoga therapists have a good understanding of how to treat only one of these bodies and have adopted a humble, loving, helpful attitude throughout their entire being, the intervention will transcend the dimension it is targeting and spread to every level. That is the secret.

Yoga Therapy for Karana Sharîra
The Causal or Spiritual Body

Karana Sharîra is the being's sacred temple, home of the individual soul, Jîva, the Divine Light within. This body is its vessel.

Yoga therapists do not treat the soul itself because it is of divine, eternal essence, one with Oneness. Its perfection is unchanging and incorruptible. However, its sheath, its energy structure, its bioelectromagnetic casing, which emanates and blends with the similar emanation from all living creatures; that structure is malleable and perfectible.

In his opus the *Tirumandiram*, siddha *Tirumûlar* mentions that the individual soul is maintained in its human adventure as long as it remains under the spell of the three *malas* (beings fundamental impurities). These malas are: *ânava* (ignorance of its true divine nature), *karma* (understood here as samskâra's generating actions) and *mâyâ* (the illusion of matter and form reality). Malas are the source of all afflictions. Psychophysical lesions are just a gross extension of the intensity of these influences. Tantras teach that through diligent Yoga practices, ones being can be freed at all levels while keeping its human physical sheath. This is when Jîva becomes Shiva.

In all cases of illness or suffering, the causal body will benefit from the Yoga therapist's adoption of a voluntary attitude, inwardly expressing *joy, compassion, love* and *equanimity*. These four qualities of love are found in classical Yoga's philosophical treatises (Patanjali, Yoga Sûtras I-33) and in Mahâyâna Buddhism, in the elegant wording of the *Four Immeasurables*.

Fully adopting this inner attitude is highly therapeutic and refines the causal body in such a way that it no longer impedes prânic flow. Remember, inner Prâna is the very expression of *Jîva*, Shakti manifested in the self.

This emanation has curative properties that Yoga therapists can use as a healing tool. This radiant glow is in fact, the shimmering of the soul. Depending on how freely it flows, clairvoyants say that its influence transcends the bounds of matter, space and time.

Therapeutic Application for Ânanda-Maya Kosha

Yoga therapy is practiced with the inner attitude corresponding to one of the Four Immeasurables or Ho'oponopono. This can be accomplished through the ritual of a daily sâdhana. If, at the beginning of the work day, the Yoga therapist regularly, fervently and enthusiastically repeats the actions that give rise to this attitude, then, equipped with this therapeutic armour, he or she can powerfully trigger the patient's self-healing.

The result is often quite wonderful. The therapist's causal body resonates with the client's and the client feels a wave of love. In this state, Prâna is set in motion in the person's entire being.

For Yoga therapists who are filled with this light, Prâna flows freely, penetrating the subtle body, the physical body and also emanating outward.

This divine flow can be intentionally directed with the power of one of the Four Immeasurables, from the Yoga therapist's heart to the patient's heart. This can be done by the laying on of hands, with or without contact, or by simple directed intention, with this inner attitude.

Silently recited prayers, mantras or a spiritual evocation can also enhance the intervention.

The approach works by its infectiousness. The person being treated enters into harmonic resonance with the therapist's vibration and is "infected" by it.

Yoga therapists who favour this type of intervention are working at the highest ethereal level of the being. The tool they are using is the most subtle and most powerful: Love.

The more that Yoga therapists master this method, the less they will need to act on the other bodies because spiritual healing, sparked by Love, has the ability to correct any problem at any level.

CHAPTER 5

Yoga Therapy for Sûkshma Sharîra
The Subtle Body

We have seen that the subtle body has three sheaths: the vital sheath, in which Prâna circulates in a privileged manner, the mental sheath (lower mind), creating thoughts, mental images and memories like an automaton, mostly in reaction to physical sensations, and the intellect (higher mind), capable of reason and discrimination, the seat of conscious awareness, volition and the ego.

Therapeutic Application for Prâna-Maya Kosha

Prâna-Maya Kosha, the vital sheath, is an exact copy of the physical body. Its Prâna circulates in every recess, expressing vitality.

Prânic fluidity emanating from Prâna-Maya Kosha animates not only the body's crude sheath, but also the sheaths of the mind. The physical body cannot be separated from its mental structure. It is an indivisible whole. We will come back to this.

Since Prâna is the essence of how all the sheaths operate, therapeutic work on Prâna-Maya Kosha will overflow into the physical and mental structures, with *Prânic healing* and Prânâyâma as the main therapeutic intervention methods.

Prânic Healing

Since Prâna-Maya Kosha is the being's prânic sheath, action can be taken on its overall clearness and fluidity so that its prâna can flow freely and spread effectively to the other sheaths.

There are several schools that teach this approach. Each follows the tenets of its founder (guru, master, guide, etc.) and features an intervention method that, invariably, calls on a more or less active laying on of hands (with or without contact) to influence the client/patient's prânic fluidity. Often, the intervention is combined with spiritual intentions.

I hope I do not offend anyone by placing these methods in the same category: *Reiki, Prânic Healing, Therapeutic Touch, Polarity,* etc.

Remarkable therapeutic results have been obtained with all of these methods, some of which have been documented in specialized literature.

Therapeutic Prânâyâmas

People have been practicing prânâyâmas, the scientific art of mastering the breath, for thousands of years. Prânâyâma yoga is certainly a good candidate for scientific investigation, being a precise practice that can be reproduced.

In the first part of the book, I mentioned that without exceptional control, humans cannot survive properly more than one month without eating solid food and more than one week without drinking. To complete this thought, we should add: *without exceptional control, humans cannot live for more than a few minutes without breathing!* Breathing can be mastered through the practice of Prânâyâma Yoga.

Knowledge of these prânâyâmas lets Yoga therapists suggest breathing exercises that can speed up or slow down metabolism, according to need. There is a wide variety of prânâyâmas to choose from: to vitalize, inhibit, raise or lower body temperature, accelerate or slow certain physiological rhythms, calm or sharpen the mind and prepare for trances in dhâranâ and dhyâna.

Mechanics of Prânâyâma

Prânâyâmas form a group of kriyâs with common characteristics. To understand how to properly carry out a prânâyâma, one first needs to grasp its structure.

> Each prânâyâma has four phases:
> 1. Inhalation or *Puraka*.
> 2. The peak of inhalation, when Puraka is suspended and the breath is held *(Antar-kumbhaka)*
> 3. Exhalation or *Recaka* (pronounced *rayshaka*).
> 4. The floor of exhalation, when Recaka is completed and the breath is held *(Bahya-kumbhaka)*.

Most of the prânâyâmas are performed through the nose. The art of Prânâyâma yoga involves exploring the different phases and various constrictions of the airflow passing through the laryngeal glottis.

Regardless of the type of prânâyâma, the respiration's prânic flow remains essentially the same. Here is an explanation:

– When Puraka (inhalation) begins, the diaphragm contracts and its dome (the central tendon of the diaphragm) drops into the thoracic cavity. This causes a depression in the lungs, which are attached to the diaphragm by the pleura.

– The ambient air outside the body, along with its Prâna, enters the upper airways (usually through the nostrils) and rushes into the lungs to fill the depression. The air column moves down.

– Prânavâyu follows the flow of incoming air and descends into the lungs from the cerebral hemispheres.

– Simultaneously, Prâna rises in Idâ nâdî, from the Mûlâdhâra cakra to the Âjnâ cakra.

There, the prânâyâma remains suspended in antar-kumbhaka. The diaphragm is in contraction and is low. The intrapulmonary and atmospheric pressure is balanced. The lungs are filled with air. Then Recaka begins:

– The diaphragm releases the contraction, its dome rising viscoelastically toward its high resting position.

— This push upward, sometimes accompanied by a contraction of the intercostal and abdominal expiratory muscles, causes pressure to rise in the lungs. The intrapulmonary air column moves upward.
— With apânavâyu, the air is expelled out of the airways.
— Simultaneously, Prâna descends along the Pingalâ nâdî, from the Âjnâ cakra to the Mûlâdhâra cakra.

There, prânâyâma can remain dormant, without movement, in bahya-kumbhaka. The diaphragm is completely relaxed and at rest in its classic high intrathoracic position. The lungs are emptied of their vital air. Then another cycle begins.

The Nasal Cycle and Swara Yoga

According to classical physiology, air flows in along the upper airways during normal respiration. At rest, this inflow is through the nostrils. The air column moves from the rhinopharynx to the pulmonary alveoli. During the expiratory process at rest, the air column is expelled again, in the opposite direction, through the nostrils.

However, one fact about respiration remains a mystery in modern physiology. The air column appears to favour inflow and outflow through one nostril rather than both in equal measure. You can check this by placing your index finger under the nostrils and exhaling in several short bursts. You will notice a more constant jet of air from one nostril than the other. Moreover, the flow of air alternates regularly between the two nostrils. In other words, we breathe more through the left nostril for a certain period of time, then the right nostril and so on, in a continuous alternating cycle.

The physiology suggests that this phenomenon is led by the autonomic nervous system, causing cyclical vasoconstriction (alternating left, right) of the arterioles of the nasopharyngeal mucosa.

Few people know that this mucosa is erectile, like the erectile structure of the genital organs. The rush of blood in the vasodilated arterioles causes the mucosa to thicken. Vasoconstriction of the same arterioles causes the mucosa to thin and the airway lumen to dilate, which leads to an increase in airflow.

There is always air circulating in both nostrils, but the density of the flow is greater in one than the other. The transition from one nostril to the other takes about 10 minutes, at which time both airways are simultaneously empty and the two nostrils expel air in equal measure. The time it takes to complete a cycle varies by individual. For the average adult at rest, respiration alternates from one nostril to the other every 60 to 90 minutes, for a complete cycle of approximately 120 to 240 minutes.

Physiology cannot explain why this alternation exists and why its time varies for a given individual and from one individual to the next.

Yogis are aware of the nasal cycle. In fact, there is an entire yogic science built around the concept. This knowledge is known as *Swara Yoga*. Ancient Sanskrit treatises were written on this topic, describing several physiological and spiritual functions:

- *Temperature control*: the nasal cycle can be used to raise or lower body temperature. Practicing specific prânâyâmas in certain poses can increase or reduce internal temperature by several degrees in a few minutes, making it a very effective therapeutic tool. Yogis noticed that certain poses favour air-flow in one nostril over the other. When in lateral decubitus, for example, airflow has a tendency to favour the part of the body facing up, the side that is free. For example, if you are lying on your right side, the air will flow mainly to the left and vice versa. When air is forced to circulate to the left, this is *Idâ prânâyâma*; it has an anti-inflammatory effect and can bring down fever in mere minutes.

- If air is forced to circulate mainly to the right, it is *Pingalâ prânâyâma*; internal temperature rises and warms the body.

- These two prânâyâmas can also be practiced in a variation of Nâdî sodhana (see the later chapter on types of prânâyâmas).

- It is said that entering samâdhi is significantly easier during the phase in which airflow is equal in both nostrils.

- The same is true for the rise of Kundalinî.

Swara Yoga uses the nasal cycle to refine the physiology and further harmonize it with the internal biological forces and the cosmic and telluric forces of nature. It teaches how to synchronize this cycle with the monthly lunar phases and how to use the alternating flow in all physical and mental activities for more efficient undertakings and optimal health. It ascribes different behaviours depending on whether respiration is to the right or left. Some authors say that parents can use this science during their union to choose the gender of their child, etc.

Respiratory Phase Proportions and Efficient Ventilation

Classic respiration for a healthy adult at rest is approximately 12 to 15 cycles per minute. That equals about 21 600 cycles every 24 hours. Yoga rishis and sages noticed that longevity is in direct proportion to respiratory rate. This is true of all living beings. The more rapid the respiration, the shorter the lifespan. With a few exceptions, healthy wild animals that breathe the slowest live the longest (more than 200 to 400 years for certain turtles). In humans, short, rapid respiration is indicative of chronic hyperventilation, the main respiratory cause of illness and physiological imbalance.

The four numbers in the ratios below, indicating the respiratory proportions to be applied, respectively refer to: ratios of inhalation time, ratio of breath retention at the peak of inhalation, ratio of exhalation time and ratio of breath retention at the end of exhalation. Moreover, all the proportions are based on the first ratio: inhalation time.

1 : 0 : 2 : 0

In seeking optimal respiratory efficiency, yogis and yoginîs through the ages began to focus on the expiratory phase of the prânâyâmas. Noting that in order to fill the lungs with air, the lungs first had to be emptied; they sought the most energetically efficient proportions for the respiratory phases. They discovered that exhalation that was twice as long as inhalation produced the best environmental extraction of Prâna. If the length of the inspiratory phase is one second, the expiratory phase should be two seconds. The rule followed here is to take the duration of the inspiratory phase and double it for the expiratory phase.

One physiological explanation for that suggested behavior is that lengthening the expiratory phase causes temporary cellular hypoxemia. Therefore, the concentration of carbon dioxide (CO_2) in the tissue increases. This hypercapnia induces bronchodilation and vasodilation, which brings more blood to the tissue (as well as Prâna) during the next inhalation.

By lengthening the time it takes to complete a respiratory cycle, not only does oxygenation and prânic intake increase, but practicing these types of prânâyâmas on a regular basis also improves life expectancy! It is now proven that regular practice gradually trains the person to breathe more slowly at rest, from the abdomen, a method that is recognized as the most efficient means of saturating the tissues with oxygen and Prâna.

The vast majority of Yoga schools around the world teach a mathematical formula for the prânâyâmas based on proportions. In my example, each number in the ratio corresponds to a reference time (usually in seconds). This reference time can obviously be adjusted according to individual capacity and is traditionally the duration of the inspiratory phase. This is the control time required to complete a maximum pulmonary inhalation at rest. In the proportional formula, it is given the value of 1. The three other numbers respectively correspond to the ratio of: retention time after inhalation, exhalation and retention time after exhalation.

A complete formula may look like this: 1 : 4 : 2 : 4. The ratio of the proposed retention after inhalation is four times as long as inhalation, followed by exhalation, which is two times as long as inhalation and finally, retention after exhalation is again four times as long as inhalation. That is a complete cycle. For a four second inhalation, the proportional formula of 1 : 4 : 2 : 4 becomes, in seconds: *4-16-8-16*.

Since some schools do not promote voluntary retention of breath, the formula would be 1 : 0 : 2 : 0 or, in seconds, *4-0-8-0*, etc. See the paragraph on kumbhaka a little further.

1 : 0 : 1 : 0

Classical respiration has a ratio of 1 : 0 : 1 : 0. If the maximum basic inhalation lasts three seconds, that is also how long the exhalation will last. This classical respiration is automatic. When we are awake, our consciousness barely participates in the process and the biomechanics are governed by the autonomic nervous system. Hormones and emotional states also play a key role in the variability of this rhythm.

There is no retention time in classical respiration.

All classic prânâyâma techniques can be practiced with the 1 : 0 : 1 : 0 proportion or the 1 : 0 : 2 : 0 proportion. The choice of ratio will be based on the suggestions made by your school. It is possible to apply a different proportion according to need or desired psychophysiological effect.

Prânâyâmas that use the 1 : 0 : 1 : 0 ratios are simpler and are often combined with a related practice. It is possible, for example, to breathe according to this ratio while taking the same number of steps as you inhale and exhale during conscious walking. Conscious walking is a Buddhist active meditation that is very effective at calming the mind and helping appreciate the present moment. During Ujjâyî breathing (see the section on the types of prânâyâmas), this proportion can be combined with a voluntary hiss of air between the vocal chords of the laryngeal glottis. This hissing sound not only focuses attention, but also clears accumulated tension and secretions from the laryngeal mucosa.

The Importance of Kumbhaka

Voluntary breath holding is a topic of special importance. Kumbhaka should only be practiced under the supervision of an experienced yogi. The physiological effects of premature retention seem to indicate a potential danger for the cardiovascular system (namely by affecting arterial blood pressure). Responsible Yoga schools advise beginners and weak or ill students to avoid retention. They often recommend that beginners practice prânâyâmas without kumbhaka for several months, even years, before modifying their practice. Some schools never teach kumbhaka at all, arguing that it will occur by itself in trance states (samâdhi) induced by prolonged practice.

I recommend that you follow the knowledgeable advice of experienced yogis and yoginîs at your school or on your path.

Prânâyâmas and Cardiac Coherence

Regulating breathing by calming it, usually through soft nasal respiration, unites and potentiates apânavâyu and prânavâyu. When it is calm and under control, respiration causes the brain and cardiovascular system to enter into a state of harmonic resonance. The heartbeat slows and stress-related physiological conditions all diminish, as demonstrated by the numerous *cardiac coherence* studies.

Cardiac Coherence

The most convincing example of respiration's immense physiological influence on the other systems is eloquently illustrated by the heart muscle's immediate reaction to controlled breathing.

By studying the heartbeat's constant and immediate variability in response to psychophysical states, scientists have discovered that the heart has a tendency to instantly enter into remarkable synchrony when the individual breathes slowly and evenly. Named *cardiac coherence*, this physiological reaction is the subject of several scientific studies, namely those conducted in the last 20 years by scientists at the *Heartmath Institute* in the United States.

The Heartmath Institute has noted so many physiological benefits when the heart is in a state of coherence that an entire teaching and stress management system was developed to promote this state. Cardiac coherence makes positive changes to the brainwaves, reduces stress hormones, lowers arterial blood pressure and improves most of the physiological functions.

The prânayâma that is the closest to the type of breathing proposed by this institute is the simplest of them all, *Sukha prânayâma*.

Preparing to Practice Prânâyâmas

Practicing prânâyâmas requires a certain preparation.

Choose a calm, unencumbered space where you will not be disturbed. Close all electronic communication devices to enjoy some peace and quiet for a time. Make yourself comfortable.

It is ideal to practice prânâyâmas with a straight spine. Experienced yogis and yoginîs can sit on the ground in one of the classic âsanas that adopt the main qualities of good posture: stability and comfort. This encourages air to flow freely in the bronchial tree. You can also sit on a chair or stool. The most important thing is to find a position that does not restrict respiration or causes the body to become a distraction (pain, discomfort, etc).

Yoga therapy clients will be guided to choose the most suitable position for them. They can even practice prânâyâmas in bed if no other âsana is possible.

Before starting, it is a good idea to release myofascial tension in the diaphragm and central axis. Very efficient diaphragm releasing techniques exist. I teach some of them in my workshops. A relaxed diaphragm is much more capable of effectively performing prânâyâmas.

All of the prânâyâmas have therapeutic benefits, although some have more noticeable, sometimes even spectacular effects.

Types of Therapeutic Prânâyâmas

If your patients wish to adopt a regular prânâyâma practice and are new to Yoga, it is a good idea to suggest that they start slowly, with a simple protocol. They will be able to refine their practice over time and lengthen the sessions or modify them by incorporating certain types of more complex prânâyâmas.

By making this important therapeutic choice, your patients will see their health and energy greatly improve.

Let us examine the practical method and the psychophysiological effects of the most common prânâyâmas:

Ujjâyî

This prânâyâma is considered the foundation of all the types of breathing exercises. Ujjâyî means *victorious respiration* or *extended breath.*

Practice:
- When applying this technique, breathe in and out slowly, narrowing the diameter of the laryngeal glottis. This voluntarily causes a hissing reminiscent of a whisper, the sound of a steam

train entering a station or even the incoming ocean tide on a quiet day. The proportion of inhalation and exhalation time is $1:0:1:0$ or $1:0:2:0$.

Psychophysiological effect:
- Concentrating on the sound of the air moving in and out deeply calms the mind and body, making the being especially receptive and efficient in all later undertakings.
- If the rhythm is $1:0:1:0$, prânâyâma can even be practiced during walking meditation, with each step counted as a beat of respiration.
- In a seated position, ujjâyî can gradually be extended to $2:0:2:0$, or $3:0:3:0$, etc. Tantric yogis believe that lengthening the ratio of respiration time is the key to longevity. Practicing this technique on a regular basis increases respiratory capacity and reduces respiratory frequency at rest.
- Ujjâyî can be added to the majority of the prânâyâmas that follow.

Sukha

Respiration of ease and gentle joy, using deep abdominal breaths, this prânâyâma is even simpler than Ujjâyî.

Practice:
- No special sound is produced during this practice. The emphasis is simply on observing the inflow and outflow of air. Breathe in and out at maximum capacity, without forcing, at a rhythm of $1:0:1:0$. Place one hand on the abdomen. The swelling of the abdomen at inhalation and deflation at exhalation will help focus the attention on the movement of the air. When the practice has been mastered, the exhalation time can be extended to $1:0:2:0$.

Psychophysiological effect:
– This simple prânâyâma reduces stress, fears and worries. It also lessens anxiety and depression. A three minute practice will bring calmness and inner peace.

Nâdî Sodhana

This very important prânâyâma is also called *Anuloma Viloma* or *alternate breathing*.

Practice:
– In this practice, the air enters one nostril and exits the other. In the classic method, sit very straight and raise your right hand in front of your face in the *Vishnu mudrâ*. Fold the index and middle finger in the palm of the right hand. Place the right thumb to the right of the right nostril, with the ring finger and little finger facing the left nostril.
– Begin by breathing calmly with both nostrils for several cycles. Then, with the thumb pressing and blocking the right nostril, breathe in, using only the left nostril.
– At the end of the inhalation, use the ring finger and little finger to block the left nostril and lift the thumb from the right nostril. The breath is therefore expelled from the right nostril.
– Keep the fingers in the same position. The inhalation begins with the right nostril. At the end of the inhalation, lift the ring finger and little finger from the left nostril and use the thumb to block the right nostril. The breath is now expelled from the left nostril. This completes one cycle.
– You can begin with ten or fifteen cycles or a ten minute session and gradually increase the number of cycles or practice time, so long as it remains comfortable. The ratio is $1:0:1:0$ or $1:0:2:0$.

- The Vishnu mudrâ can be replaced with the *Nasagra mudrâ*. The difference is that the index and middle fingers press on the area between the eyebrows, the sixth cakra, Âjnâ. Or, the nostrils can simply be blocked with the thumb and index fingers placed on either side (like a clothespin). The important thing is to create a cycle that alternates nostrils.

Psychophysiological effect:
- Practicing Nâdî sodhana regularly for a few months is said to cleanse the subtle body's nâdîs, making it even more fluid and radiant.
- The nostrils are connected to the opposite sides of the contralateral cerebral hemisphere. Breathing through each nostril therefore harmonizes the right and left sides of the brain and the autonomic nervous system.
- The direct effect on the body is a lower heart rate, reduced stress and anxiety, an increase in general vitality as well as improved memory and concentration.
- This practice will also lengthen the transition time in the classic physiological alternation of the nasal cycle and equalize the time in each nostril. The transition period, when air circulates equally in both nostrils, is said to be especially conducive to spiritual activities. Nâdî sodhana will balance prânic flow on both sides of the vital sheath, especially in Idâ and Pingalâ.
- There are several possible variations to this prânâyâma. One is to only breathe through one nostril for an entire session. This is called *Idâ prânâyâma* or *Chandra prânâyâma* (left side, Moon respiration) or *Pingalâ prânâyâma* or *Surya prânâyâma* (right side, Sun respiration). These will have the same temperature controlling effects mentioned earlier in the section on Swara Yoga.

Shîtalî

Respiration of calmness and appeasement. Shîtal means calmness stemming from coolness. It is one of the rare prânâyâmas to be practiced with the mouth.

Practice:
- In stable, comfortable, seated position, back straight, roll your tongue so that both sides meet at the top, creating a tube. Extend the rolled tongue outside the mouth and place your lips around the tube. This will help maintain the tube shape and prevent air from circulating anywhere but in the tube. Breathe in slowly through the tube, all the way to the back of the throat. You will feel immediate coolness. At the end of this long inhalation, pull your tongue back in, close your mouth and exhale softly through your nose.
- Note that the ability to roll the tongue lengthwise is a genetic trait not shared by all. Some individuals will therefore not be able to do it. However, they can still practice Shîtalî by simply extending their tongue and resting it on the lower lip. The inhaled air will pass over the tongue and produce a similar effect. Exhale the same way. This variation is called *Sîtakârî prânâyâma.*

Psychophysiological effect:
- This is a cooling respiration. It eases mental activity by refreshing the mind and body. When internal temperature drops, so does fever. The body becomes less acidic. Stress, anger and rage are regulated. This prânâyâma brings about a sense of calmness, tranquility and peace.
- It also eases feverishness and insomnia.
- Repeat as many times as needed to obtain internal coolness.
- Do not perform this exercise when you have a cold or flu.

Bhastrikâ

Bellows respiration gets its name from the sound that is produced during this practice, a sound similar to fire bellows, such as those used by blacksmiths.

Practice:
- This is a powerful respiration, as active during inhalation as at exhalation. It is a good idea to blow and clear your nose before starting, otherwise the practice will do it for you!
- Seated in a comfortable, upright position, breathe in and completely fill the lungs so that the thorax lifts to the upper ribs. Then, exhale forcefully through the nose, from the top down. The force of expiration is from the shoulders to the diaphragm, toward the belly button, ending with an abdominal contraction that expels the remaining air by actively pushing it out. The next inhalation begins immediately with a new inflation of the lungs, with the thorax actively lifting the ribcage and shoulders. One cycle is about five breaths. After completing a cycle, breathe softly once or twice, and then start another cycle. Start slowly. With time, the cycles will shorten and become more rapid and intense.
- One variation, depending on the school, is to use the same respiration, but lift the upper limbs over the head when breathing in and lower them vigorously, elbows and fists tight to the torso, when breathing out.
- Another variation suggests exhaling through the mouth with pursed lips.
- Do two cycles per day for several weeks, then gradually increase to four or five cycles per day. When you have mastered this prânâyâma, you will be able to do the respiration more quickly (2 per second or more), with as much energy.

- You can also explore that prânâyâma by breathing from a single nostril as with Nâdî sodhana.

Psychophysiological effect:

- Some texts mention a prânâyâma that heals illness when the practitioner visualizes waste products and negative energy being expelled from the body during forceful exhalation. Bhastrikâ speeds up the basal metabolic rate and therefore burns more calories. It purifies both the physical and subtle body by helping eliminate toxins and waste. It clears and strengthens the lungs and respiratory system and fortifies the nervous and digestive systems. It gives a refreshing feeling of joy. It awakens Kundalinî.
- This prânâyâma is contraindicated for most heart conditions, hypertension, aneurysm, the risk of stroke and epilepsy.
- Stop immediately if you experience any discomfort, pain, nausea or dizziness.

Kapâla-Bhâti

Called *skull illumination respiration*, this prânâyâma is also one of the six Shatkarmas practices, Hatha Yoga's classic physical purification exercises.

Kapâla means skull, *Bhâti* means light. This practice is said to fill the mind with light.

This prânâyâma is more rapid than Bhastrikâ. It is called the skull illumination prânâyâma because when it is practiced on a regular basis, the vital centre between the eyebrows (sixth cakra, Âjnâ) is said to emit a calming internal light.

Practice:

- Sit in a comfortable, upright position. Inhale slowly to two thirds of your capacity, then, with a sudden abdominal contraction, perform a series of short, powerful, forced exhalations from the nostrils, as if you are trying to clear your sinuses.

- In this prânâyâma, emphasis is placed on exhalation. Forced evacuation occurs from the bottom up, with the lower abdominal muscles pushing the air up and out. After each exhalation, inhalation occurs automatically as the diaphragm and abdominal wall spring back viscoelastically.
- Some schools teach exhaling through pursed lips. This method is without a doubt a more powerful exhalation.
- Apply the method that feels the best. Both methods are efficient, but for different reasons (see *psychophysiological effect* below).
- Begin with a 10-15 respiration cycle and gradually increase to approximately 25 repetitions. After each cycle, take a few soft breaths, and then continue. Start with three cycles and gradually work up to five.

Psychophysiological effect:
- This prânâyâma clears the respiratory system and cleans a lot of toxins. It balances the inner pH. It stimulates brain activity and awakening. It balances and fortifies the nervous system. It improves digestion, strengthens the abdominal muscles, the diaphragm and heart. It improves blood circulation, especially venous return, by its effect on the diaphragm.
- The pursed lips exhalation method permits a copious toxins and waste disposal through its important humidity elimination (mouth cavity is more humid than nasal cavity). It also permits a strong disposal of excess mucus from the bronchial tree. But in doing so, the viscoelastic recoil of the diaphragm and thoracic cage induce an automatic inhalation through the mouth and so dries it up way faster than through the nose breathing.

- The usual exhalation method of Kapâla-Bhâti is through the nose. In doing so, the airflow stimulates the mucosa lining of the nasal cavity's roof. The mucosa is filled with olfactory receptors and so offers a direct stimulation pathway towards the brain's frontal and temporal lobes. That could be an explanation for the euphoria and light-headedness given by this practice. Here the nasal exhalation permits a more efficient toxin and excess mucous disposal from the sinuses.
- Kapâla-Bhâti is contraindicated for most heart conditions, hypertension, aneurysm, the risk of cerebrovascular accident (CVA or stroke) and epilepsy.

These two vigorous practices (Bhastrikâ and Kapâla-Bhâti) can be used to clear the organism not only of physical impurities, but also psycho-emotional negativity. Visualizing their expulsion from the body can be used very effectively with both of these prânâyâmas.

Breath of Fire or Prânâyâma of Fire

This prânâyâma is mainly performed by practitioners of Kundalinî Yoga, but will benefit everyone!

Practice:
- This prânâyâma is characterized by a series of very short, albeit quite powerful nasal respirations, with the focus placed on fluidity.
- Breathe in to two thirds of your capacity, and then do a series of small, halting respirations. Inhale and exhale the same length of time, but focus on the end of the nose or the entrance of the nostrils. The effect is a quick sniff.
- *Beginner students are often taught to visualize the respiratory action of a dog panting during a hot day.*
- Gradually increase the length of the practice, from 30 seconds to three minutes or more.

- It may be difficult to synchronize the rhythm to make this exercise fluid. With time, you will be able to sniff more quickly and powerfully.

Psychophysiological effect:
- The immediate result is intense abdominal heat, along with an inner sense of joy, freedom or expansion.
- This prânâyâma raises body temperature and harmonizes the visceral and glandular functions.
- Breath of Fire is known to open or clear the nâdîs. Prâna is therefore able to flow more freely. This multiplies the psychophysiological effects of all subsequent practices.
- This prânâyâma is contraindicated for most heart conditions, hypertension, aneurysm, the risk of stroke and epilepsy.
- Stop immediately if you experience any discomfort, pain, nausea or dizziness.

A certain euphoria may be felt when practicing Bhastrikâ, Kapâla-Bhâti and Breath of Fire. This is due to intense, temporary oxygenation and hypocapnia (drop in tissue concentration of CO_2). The resulting hyperventilation is temporary and should not be dangerous, resorbing quite rapidly when the level of carbon dioxyde returns to normal.

Brâhmarî

Bee buzzing respiration

Practice:
- Sit in a comfortable and stable position, back straight. Make a sound as you inhale through the nose, the same sound you would make if you were startled, but not quite as violent. Make a similar sound when exhaling. The result is a sound like a bee buzzing from flower to flower. The tone is often higher-pitched at inhalation. The ratio is 1 : 0 : 2 : 0 or 1 : 0 : 3 : 0.

- This prânâyâma is even more powerful if you concentrate on the sound vibration produced when you block the external auditory canal. This is done with the *Shanmukhi mudrâ*, using the fingers to block the ears, eyes, nostrils and mouth. The nostrils are only covered in this case, not completely blocked, as you need to breathe for this prânâyâma. You can also use earplugs. The sound vibration is greatly increased, flooding your consciousness. It can be made to travel within the bone structure of the skull and the rest of the body.
- Concentrate first on the vertex of the skull, gradually feeling the sound wave flow over it, and then make it descend the spinal column all the way to the coccyx. Then, let the vibration spread to every corner of the body so that all the cells are submerged.
- Begin with five repetitions, gradually working up to ten or fifteen per session.

Psychophysiological effect:
- This prânâyâma calms general tension and anxiety. It dissolves anger. By calming, it lowers blood pressure. It relieves insomnia. It clears and strengthens the vocal cords while alleviating throat problems. It stimulates better digestion and blood flow and strengthens the abdominal girdle. It produces a mellow joy that is conducive to a meditative state and helps actively awaken Kundalinî.

Matrikâ

Respiration of health, energy and dynamism

Practice:
- This is a simple respiration. Inhale and exhale with a hissing sound as in Ujjâyî, without forcing too much, at a rhythm that is twice your maximal inhalation at rest, at a ratio of

2 : 0 : 2 : 0. This prânâyâma boosts the immune system, provides energy when you are tired and gives power to those already in good health. Do 15 to 25 cycles.

Psychophysiological effect:
– This respiration often helps raise energy levels and starts your patients on the path back to health.

Sleeping Prânâyâma

This therapeutic prânâyâma is a brilliant answer to insomnia.

There are several types of insomnia, with many causes, but the common characteristic is that the agitated mind will not quiet down enough to let you fall sleep.

Practice:
– Ideally, this prânâyâma should be practiced while lying down, to more easily induce sleep.
– Lay on your back in the classic savâsana position (*dorsal decubitus*). With your eyes closed, take a long, soft breath through your nose, followed by a shorter, unforced exhalation, like a sigh (still through your nose). The ratio is 2 : 0 : 1 : 0. Mentally count about 15 cycles, no more. This should be sufficient to calm the mind and body. Often, you will not even reach 15 cycles before falling into peaceful sleep. If you are still unable to sleep, you can repeat the exercise.

Psychophysiological effect:
– This prânâyâma is very effective at relaxing the mind and body by inhibiting the sympathetic nervous system. It slows the heart rate, respiration and all metabolic processes.
– This prânâyâma is obviously contraindicated in situations requiring sustained attention or vigorous physical activity as well as for people suffering from hypotension or anemia.
– Sleeping Prânâyâma is also effective for brief, deep relaxation (such as a *power nap*).

Important Considerations

Several schools teach variations of most of these prânâyâmas together with kumbhaka. It is not necessary to voluntarily retain the breath during these exercises in order to reap the benefits, which is why I do not insist on it. If you want to explore this practice, I recommend that you consult an experienced yogi in your school or on your path.

Even with the explanations provided in this book (or any other), nothing equals practice supervised by someone with experience, such as a Yoga master or teacher.

Drishti, Bandhas, Mudrâs and Prânâyâmas

All of these previous respiratory practices will result in greater concentration of attention if they are performed with *drishti*, specific focalised gaze.

Most of the prânâyâmas can be practiced effectively with the *Shâmbavî mudrâ* (seal of Shiva), in which the practitioner focuses between the eyebrows, inside the skull.

Practitioners are often encouraged to adopt the *Jnâna mudrâ*, pressing the thumb and index finger together in an open hand, with the back of the hand resting on crossed knees, in the lotus or siddhâsana position.

Some prânâyâmas can also be practiced with Mûlabandha or Mahâbandha.

Through a regular prânâyâma practice, Yoga therapy patient will have their prânic fluidity potentiated and circulating powerfully in their being. Prâna-Maya Kosha will vitalize Anna-Maya Kosha and also flow into its mental dimension, the mind.

The mind and body is an indivisible unit. Any actions taken on one will inevitable affect the other. We know that mental activity affects the respiratory rate and that the respiratory rate affects the mind by calming it or stimulating it. Prânâyâmas are therefore a tool of choice for inner appeasement, a useful foundation for more advanced therapeutic practices aiming to quiet the mind.

Therapeutic Application for Mano-Maya Kosha and Vijnâna-Maya Kosha

When Prâna circulates in the mind, it activates its Mano-Maya Kosha. The lower mind creates images and concepts when triggered by the senses. This stimulation can become so intense that it can fill the entire mental space with unrestrained activity. The activity remains unconscious from its source, operating at the subconscious level, and the person can stay in this more or less agitated state for quite a long period of time. The duration may vary according to the intensity of the stimulation and the individual's level of mental mastery. It can sometimes last the person's entire lifetime. Inner agitation is a source of confusion and often becomes a fertile ground for chronic pathological manifestations. The solution is to quiet the mind, which is not always easy.

The mind will respond to different treatment parameters: in addition to prânâyâmas that calm the two mental sheaths, the most effective kriyâs for this are dhâranâ (concentration), and dhyâna (meditation). It is Vijnâna-Maya Kosha, Buddhi, which will consciously start this work.

Dhârana

Contrary to the classic understanding of the concept, concentration or dhâranâ is essentially a physical activity that extends to the mind. Often, dhâranâ will be the prelude, the introduction, to dhyâna. As a solitary practice, there are external kriyâs like trâtaka, in which the practitioner focuses on the flame of a candle, or other similar exercises. In practice, this means that you are beginning dhâranâ via the body. Here is how.

In a quiet setting, free from external sensory distractions get comfortable for the sâdhana. For optimal prânic fluidity, keep the spine straight without forcing. A backrest can be very helpful or a comfortable chair. Supine position should be avoided because it can lead to sleep. If the person is bedridden, this practice is still possible, although greater care must be taken not to succumb to drowsiness.

Once settled in a stable and comfortable position, with the body completely still, close your eyes and go within. This is *pratyâhâra*, the beginning of the physical withdrawal of the senses. In this phase of concentration, physical sensations gradually disappear and the mind begins to focus, usually on a body part or external object that has been internalized. Often, the mental activity will be led within by one of the inner senses.

The inner senses are subtle mirrors of the physical senses. Their mental activation, their mode of action, is triggered by memory. Sight becomes *visualization*, hearing becomes *internal listening*, touch becomes *tactile internal sensation*, smell becomes the *memory of fragrances* and taste becomes the *memory of flavours*. When concentrating in a still body without movement, the mind begins to favour one of these senses. The sense of visualization appears to be the most easily accessible.

It is Buddhi, the intellectual sheath of Sûkshma Sharîra, which consciously produces this concentration. It draws internalized images or sensations needed for focalization from the Manas memory bank.

There are two possibilities at this level: concentrating with or without object.

Dhâranâ with Object

The object (flower, mountain, fire, etc.) becomes the theme of the dhâranâ. The person fills his mental space by visualizing this one thing, or concentrating on hearing it, touching it, smelling it or tasting it. The goal is not to have a discourse on the object, but only to set it in your mind with an internal sense. Nothing else will be able to disturb this space. The object can also be an inspiring person, a divinity, etc.

Suppose you concentrate on a flower. The flower is visualized in its entirety through complete immersion.

Dhâranâ without Object

This variation involves the intense observation of a concept (love, action, fullness, emptiness, abundance, etc.). Regardless of the theme, the mental space is filled with this abstract idea by way of internal sensory observation. Again, the goal is not inner discourse, only the attentive presence of the observing mind.

The Dhâranâ Process

Whether you are practicing with or without an object, the dhâranâ remains the same: a gradual, conscious physical desensitization by concentrating on a given object or abstract concept. There is no emotion involved, only focalized attention, on a flower, for example. Your body quickly loses contact with the physical senses. It becomes numb and you even stop feeling it, you are in dhâranâ. You lose all sense of time and can remain submerged in this experience for quite a long period. The flower or other object fills your entire experiential space. That is when you enter dhyâna.

Advantages

When you master dhâranâ, your mind becomes amazingly sharp. With regular practice, you will be able to use dhâranâ to quickly grasp the essence of the events in your life. This can become an important source of awareness and creative inspiration.

When experiencing pain or illness, concentration brings a clear view of the condition and helps trigger healing.

Yoga therapists can teach dhâranâ to their clients. They can also use it to get more information about a client's state of health.

Dhyâna

Dhyâna is experienced entirely in Buddhi, the intellectual sheath. The physical body has disappeared in pratyâhâra.

Dhyâna usually occurs as an extension of dhârânâ. When the latter is prolonged, the upper mind enters into a more complete state of absorption.

The main difference is the dynamics involved. In dhâranâ, the mind focuses on a given object with piercing observation stripped of discourse. In dhyâna, the mechanism is no longer concentration, but the fluidity of the experience. The practitioner can explore the subject with every sense and the entire intellect. The ideas, images and concepts that arise relate directly to this object of concentration and orbit around it, defining it even further. This time, if there is a discourse, it concerns the object. All other discursive forms are set aside and gently eliminated from the process.

The meditation can be extremely refined and focus on emptiness or any other basic abstract theme. Some schools suggest a meditation in which the aim is to empty the mind of all concepts while remaining paradoxically alert. As in dhâranâ, dhyâna could be done with or without an object.

The meditative experience brings calmness and inner peace. Lucidity is intensified even though you are no longer aware of the outer world. These qualities carry over into daily life after the meditation practice has ended.

No matter its form or school of origin, meditation's positive impacts on physical, mental and spiritual health are now widely accepted and recognized. Innumerable studies have been conducted on these benefits, all schools combined.

Every religion and philosophy throughout history has developed meditation techniques.

Yoga therapists can teach some of the simpler dhyâna techniques, bringing about remarkable therapeutic transformations.

However, they should first practice dhyâna on themselves before using it in therapy.

Notion of Time

Several commentators mention that concentration and meditation practices require a considerable investment of time. Some say that these concrete activities require several years of intense asceticism before a light is seen at the mental horizon. That is not quite true. It is possible, in a few days of practice, to achieve a remarkable depth of concentration and meditation.

I will not deny that it often takes years to master this practice. This is true for most individuals. With some rare exceptions, mastery is usually based on experience (which, naturally, is acquired over time). Of course, there are people who will be more adept than others right from the start. But persistence always yields results. If you practice these kriyâs diligently, you will experience transformative effects in a few days, regardless of where you start on the road to consciousness!

It is typically suggested that meditation sessions last a minimum of 15 to 30 minutes, ideally morning and evening, ideally around the same time. Some schools suggest practicing prânâyâmas and meditation before the morning's physical exercises to gently prepare the body for the more intense practice of âsanas. Other schools favour starting with âsanas, arguing that this makes the body more relaxed and concentrated in preparation for meditation. I personally think that both arguments are equally valid. The important thing is to practice!

You can enter directly into a meditative state or do so gradually, by applying pratyâhâra, dhâranâ and dhyâna in succession. Since dhyâna is a glorified form of concentration, it is suggested that you spend more time on this than on concentration.

One last consideration about the notion of time... Moving from pratyâhâra to a state of concentration, to a meditative state is not necessarily a linear process. In ideal environmental conditions and the right inner disposition, it may only take a few seconds! For example, the yogi closes his eyes and pratyâhâra is achieved in the first breath. In the second, dhâranâ is established. In the third, dhyâna is already developing! The process is not linear, somewhat like the sap in a plant circulates everywhere at once or the blood in the body circulates everywhere at once. Achieving meditation can be immediate, without prior preparation. In fact, it is possible to enter dhyâna (or dhâranâ) spontaneously.

Samâdhi

Due to the way the body-mind operates, while most behaviour and experiences occur at a subconscious level and while concentration and meditation are more of a conscious experience, *samâdhi* causes us to enter ecstasy in *superconsciousness* (or rather *enstasy*, as suggested by Mircea Éliade, since it is actually an internal absorption).

A certain distance remains in dhâranâ and dhyâna, a subtle separation between the person observing and the object being observed. This distance is dissolved in samâdhi.

It is quite rare for people to experience samâdhi in their daily lives. However, it remains one of the goals of all yogic practices.

Samâdhi merges the observer and the observed, erasing the distinction between subject and object. The only thing that remains is the unified subject, which expresses itself as such. The object becomes the subject.

Commentators who have experienced this state say that it has several levels. As in meditation, the outer world disappears, leaving only the new subject-object unity.

This extraordinary identification is always accompanied by a very keen state of alertness, a feeling of bliss and various degrees of inner fullness, depending on the level achieved in this ecstatic union.

No one should judge another person's inner experience. It is extremely difficult to assess the value and progress of this type of introspection.

However, I believe that anyone who can explore his or her consciousness up to samâdhi possesses a therapeutic tool of infinite power. Nature yields completely to a being's willingness to heal when in this state. That is when miracles happen.

Samyama and Prajnâ

The practice of dhâranâ, dhyâna and samâdhi on a same object is called *samyama*. Patanjali talks about this in the third part of his *Yoga Sûtras*. Few schools teach this or know how to. Achieving samyama is as rare as samâdhi itself. When samyama is achieved on an object, its full nature is revealed. Complete, freeing knowledge of the object arises. The texts say that mastering this science results in clear-sightedness, ultimate wisdom: *prajnâ*. If the object is illness, or even better, healing, nothing is impossible for the inquisitive mind rooted in prajnâ. The Yoga therapist can then become a healer…

This is quite a bold undertaking. It will not be easily accessible to all therapists, given that it is an advanced Yoga practice. However, since Yoga therapists are also yogis and yoginîs, they can make acquiring this ability part of their art and science. The more that therapists are engaged in the action, the more they will believe in it, the closer they will come to samâdhi, samyama and therefore prajnâ.

When we are imbued with humility and power in an inner attitude that invites healing (by reaching the appropriate states in Kârana Sharîra), knows how to make Prâna flow and focus our mind on serenity (by inner work on Sûkshma Sharîra), we can transform matter altered by illness. We are able to act on Sthula Sharîra, the gross body.

Yoga Therapy for Sthula Sharîra
The Gross (Coarse) Body

Sthula Sharîra's physical sheath is Anna-Maya Kosha. In Yoga, Anna-Maya Kosha's evolution is essentially based on two behavioural patterns: ingesting nutrients and physical activity. But fundamentally, these two activities produce the same result: improved harnessing and use of Prâna.

In the first part, we saw that a yogic diet (capable of supplying Prâna abundantly and fluidly), favours fresh food, the least processed possible, grown locally, prepared with love, etc., along with healthy hydration. This diet is a therapy in itself when Yoga therapists need to support their clients in shifting to healthier eating habits.

We saw that Hatha Yoga is a form of Tantra Yoga that was specifically developed to promote increased vitality (the *best prânic fluidity*) in this physical sheath. In this sense, Hatha Yoga encompasses Prânâyâma Yoga.

These two paths will therefore be the Yoga therapist's initial tools for treating physical problems.

However, before suggesting dietary approaches or âsana and prânâyâma sequences, you should understand that clients-patients consult a Yoga therapist in the hopes of righting some imbalance or dysfunction. This physical dysfunction is accompanied by symptoms (inflammation, edema, etc.) that should ideally be alleviated after the intervention.

The main aim of the Yoga therapy consultation is not learning yogic techniques. These will be the means to healing, not the goal of the intervention. In fact, the primary reason for consulting will most often be *pain*.

Pain, Suffering and Yoga Therapy

Most people consult a therapist for a range of problems, all with one common characteristic: pain. People will consult you, the Yoga therapist, to alleviate their pain. In their mind, pain is essentially a physical experience. They can specify the type of pain, its intensity, frequency and location. Its cause is sometimes traumatic, sometimes unknown. It can be acute or chronic.

The International Association for the Study of Pain (IASP) defines pain as follows:

"An unpleasant sensory and emotional experience associated with actual or potential tissue damage, as described in terms of such damage."

Sensory Receptors

Pain is a sensation, a physical and emotional experience. It is caused by overstimulation of the sensory receptors.

A receptor is an organ of the nervous system. There are several types. These are ultra-specialized sensory organs that usually react to a precise type of stimulus and not to others. They are found in various concentrations throughout the physical sheath, especially in the connective tissue. It is for this reason that researchers currently studying the fascias are beginning to view the connective tissue in its entirety, as a massive sensory organ. In this sense, the connective tissue is the body's most extensive and voluminous organ, far larger than the skin or liver.

When a receptor is stimulated, it sends the brain afferent information by way of a nerve impulse travelling along the neural pathway. That is its classic physiology. Each receptor can trigger information perceived as painful when its maximal threshold of sensitivity is exceeded by overly intense stimuli.

Brain and Mind

The physical sheath, with its nervous system and especially the brain, forms the key instrument of the mind (Mano-Maya Kosha and Vijnâna-Maya Kosha), the optimal interface with the mental sphere. The brain integrates the impulses while the mind identifies and interprets the information. This interpretation dictates the chain of events, the motor reaction (reflex or not) appropriate to the sensation.

An afferent sensory impulse (electrochemical in nature) whose intensity exceeds the physiological threshold of comfort is interpreted by the mind as the opposite of comfort, or pain. As long as this stimulus is present, the experience will be interpreted as painful.

Genesis of Suffering

We saw, in Chapter 2, how sensation leads to emotion.

Emotion is an inner state triggered by the almost always unconscious interpretation of a chain of events whose physical origin is rooted in the senses. Emotions can be positive, generating well-being and favouring health and growth. They can also be negative, helping create or maintain a morbid process.

Emotions produce a vibration that is highly communicable in the immediate environment. A person already sensitive to this vibration can enter into resonance with it, feeling the same emotion emitted by the other person, without being aware of its source. This ability can be harmful for people who are unaware of what is happening. However, it can be a powerful therapeutic tool for those who know how to control this state.

Pain can generate one or several emotions in the body-mind, creating a reverberating sensory circuit, a vicious cycle that further heightens the painful experience. Ultimately, if the pain stimulus is intensified, the mind will cut contact with the physical body and the person will lose consciousness.

When pain becomes chronic, it almost always gives rise to one or more negative emotions (grief, sorrow, anger, despondency, guilt, frustration, victimization, etc,). And then, in the body-mind, pain evolves into suffering.

It is also possible for suffering to manifest without prior physical pain. In fact, relatively frequently, suffering is the morbid evolution of a negative emotional state, a form of perfidious discontent, chronically eating away at the person's mind.

Suffering is a profound experience. It gradually disrupts the person's vital ground and ultimately destroys both the body and mind. It can manifest to the being through all the known variants of depression and neurotic anxiety disorders and can even lead to psychosis. It can be insidious and subconscious, progressing slowly, or all at once, arriving in a flood. Suffering is stress that causes illness in the mind-body

Pain does not kill, but suffering does. Someone who is suffering will seek relief, whether consciously or not, a release that sometimes only comes with death. Suffering can destroy people. Despair sometimes leads to suicide, a final solution to a temporary condition.

People can suffer for a long time. They can also manage to adapt to suffering, to transcend it or sublimate it. Suffering can be a remarkable teacher, imparting countless life lessons. But it can also be overcome.

Solutions

In orthodox medicine, pain is relieved through pharmacodynamics or surgery. This approach is only relevant if gentle, alternative methods have not yielded the hoped-for results. The methods that are used often present greater risks than the illness itself. The side effects of analgesic, anti-inflammatory or psychotic drugs are sometimes as serious as the symptoms themselves and, all too often, only treat the physical aspect of the morbid conditions.

However, the relevance of taking psychotic drugs for major depression greatly exceeds the inconvenience of the side effects. They should not be dismissed out of hand. Combining psychotherapy with medication and a balanced diet, along with regular physical activity, will effectively support a return to health. Gradually, under the care of a doctor and various caregivers, dependency on the medication can be reduced. People suffering from such afflictions should be carefully guided in their search for solutions.

Applying the Yoga therapy protocol will immediately reveal its simplicity and appropriateness.

Yoga Therapy Protocol for Pain

Even if the pain is mainly physical, the three bodies are actively involved in its resolution.

Sthula Sharîra, gross body:
Âsana: Find a comfortable position in each yogâsana, one that is the least painful. Avoid all aggravating positions.
Prânâyâma: Sukha, Ujjâyî, Matrikâ prânâyâma breathing
Pratyâhâra: Voluntary withdrawal of the senses.
Drishti: Direct the focus to the third eye, with the eyes closed.
Sûkshma Sharîra, subtle body:
Dhâranâ: Concentrate on the body part concerned, Anapana-sati.
Dhyâna: Dissolution meditation (Transcendental Meditation, Samatha, Shuddi Dhyâna Kriyâ, etc.).
Kârana Sharîra, causal body:
Samâdhi or ecstatic contemplation.
Adopt the inner attitude of the Four Immeasurables.
Ho'oponopono.
Any other elevating or transcending inner activity.

Comments

The therapy requires space and time conducive to its application.
   – Set aside 30 minutes per day, preferably at the same time. Your state of ill health is active day and night, seven days a week.

Your body-mind is also fighting this state on a full-time basis, often without you truly being aware of it. Thirty minutes will be your conscious grain of salt in the process. It will greatly increase your self-healing power.

– Do the exercises on an empty stomach. Empty your bladder and void your bowels.
– Create a quiet space, making sure that life's daily hustle and bustle cannot penetrate your temple of peace. Shut off the telephone, radio, television and computer. You want to avoid all mental distractions that could scatter your thoughts or pull you from your intention to restore well-being. This space can be outside, if your environment permits.
– Your isolation can be supported by soft background music. Know, however, that silence is a very effective therapeutic tool. Someone who subscribes to your approach can also assist you.

The protocol is ultimately applied in the three dimensions of the being simultaneously. In the beginning, however, it may be useful to start by establishing the process in the causal body (Karana Sharîra). If you can achieve a state of benevolence, compassion, joy or equanimity in your soul or apply Ho'oponopono, you will be able to more easily free your body-mind from pain.

The aim is not to deny your pain, but to accept it for what it is: a sensation interpreted by the mind. Regardless of its initial cause, you need to take full responsibility for it, because your body-mind has claimed it. Victimization and self-pity will only aggravate the pain. They are useless and counter-productive.

Find your âsana, which should be stable and comfortable. To maximize prânic fluidity, it is preferable to keep the spine straight. However, if this causes great pain or discomfort, lying down is acceptable.

In a gentle, continual, fluid and deep prânâyâma love yourself completely as you are in this moment. Feel a wave of compassion wash over you.

Dhârana begins with identifying the body part that is affected. This concentration is accompanied by pratyâhâra, a gradual withdrawal from bodily sensations, like a slowly spreading anesthetic. When you are deep in dhârana, your concentration is such that your body is completely immobile and sensations have almost fully disappeared. This state is sometimes difficult to achieve if you are not in a quiet environment conducive to concentration. Do not get discouraged. Continue with your practice and success will yield great rewards.

Dhyâna continues the process with gradual and complete mental dissolution. This meditation exists in all philosophical schools, whether in Buddhism's Samatha, Kriyâ Yoga's Shuddi or any other form of emptiness meditation.

Continue to firmly bring your wandering mind back to these different parameters.

Repeat the session during the day, if so desired.

You will be astonished to experience a significant reduction in all types of pain in a few days, often with complete relief.

Impermanence

One of Buddhism's great lessons about suffering or pain is that all conditions, positive or negative, are impermanent. All conditions come to an end. Contemplating impermanence is one of the Buddhist meditation practices (Vipassana) and is itself a perfectly appropriate solution to suffering.

Fatigue and Lack of Energy

One of the characteristics of the morbid process in the body-mind is that illness is like a vampire, feeding on the host's vitality. The person, often unconsciously, gives the state of illness so much energy that he or she soon experiences a flagging, a lack of vigour expressed as great fatigue. This listlessness is often exacerbated by the fact that the inner forces fighting the illness are drawing from the same reserves of vital energy. That is why chronic illness is so debilitating. The body-mind is robbed of its vitality from two opposing sides. If the illness is stronger, it takes all the vitality and kills the host. If the healing is stronger, it destroys the illness and feeds the host. It all depends on which side wins the fight, the morbid state or the person's own capacity for self-healing. The Yoga therapist's goal is to tip the scales toward healing.

Matrikâ prânâyâma, the energy prânâyâma (2 : 0 : 2 : 0) is a wonderful tool, providing support and encouraging the restoration of health and energy. It is a must.

Stress and Structure

Anything that affects the integrity of the structure is bound to spread to all levels of the being. A physical injury can be the start of a morbid state in the other sheaths. A moral injury can turn into a physical ailment.

In all cases, an injury is a type of stress, uncommon tension that can produce physical or subtle damage. The presence of such tension changes the body's biochemistry, causing it to release abnormally high levels of cortisol and adrenaline (stress hormones) in the bloodstream. A rise in these hormone levels disrupts the entire endocrine system and contributes to the acidification of the inner environment.

The notion of stress is now a key topic in the genesis of pathology. Whether its origin is physical, vital or mental, every time *eustress* (stress that is acceptable, healthy and necessary for maintaining health) is surpassed, the environment becomes fertile grounds for developing a lesional state.

Therapists have long noted, in a clinical setting, that a person who is ill or injured does not have the same structural flexibility as someone in good health. In the physical sheath, the latest research on the fascias shows that when under constant stress, the cells of the connective tissues (*fibroblasts*) tend to become *myofibroblasts* (contractile cells). This would explain the contractility and retraction of the injured tissue as well as the lack of flexibility seen in injured or ill people.

Antistress Yoga Therapy Protocol

For people living in this society, at this time in history, it is impossible to avoid stress. To counter its negative effects on the being, we can again turn to Yoga, which offers several ways to cope:

Sthula Sharîra, gross body:
Âsana: Regular practice of a sequence of âsanas.
Prânâyâma: Sukha, Ujjâyî, Matrikâ, Bhastrikâ, Kapâla-bhâti, etc.
Physical activity: The secret is regular practice, at least three times per week (walking, hiking, running, cycling, swimming, team sports, etc.).

Sûkshma Sharîra, subtle body:
Dhâranâ: Concentration leading to dhyâna, Anapana-sati.
Dhyâna: Dissolution meditation, Samatha or equivalent.

Kârana Sharîra, causal body:
Samâdhi or ecstatic contemplation.
Adopting an inner attitude of the Four Immeasurables.
Ho'oponopono.
Any other elevating or transcending inner activity.

Acute and Chronic Injury or Illness

In the physical body, managing impairments will vary based on their chronicity.

An injury is acute when it has just occurred, within a few hours or days, usually less than a week. It is subacute when it lasts less than three months. It is chronic when it persists more than three months. It is not uncommon for a chronic injury to last several years.

Most illnesses are chronic by nature.

Holistic Management of Injuries and Illnesses

Before taking action, Yoga therapists need to be in an inner state that favours healing. This is achieved by centering on Karana Sharîra with the appropriate attitude conducive to the intervention (joy, compassion, love, equanimity, Ho'oponopono, etc.). It may also be useful to complete this preparation by flooding Sûkshma Sharîra, the subtle body, with radiance and light in a meditation and prânâyâma session. A daily sâdhana ensures continuity. In doing this, Yoga therapists don their armour as healers, ready to take on any pathology.

A physical injury, regardless of its origin and extent, informs the Yoga therapist that prânic fluidity has been disrupted.

The state of injury (acute or chronic) in a tissue or in the organism expresses this loss of fluidity as:
  – local or overall inflammation
  – reduced range of motion
  – loss of strength
  – inhibited function
  – restricted working capacity

The proposed treatment aims to control inflammation and restore mobility (range of motion, strength and energy), a sign of vitality's return.

Yoga therapy's holistic approach allows the therapist to view the problem as a whole. A lesion can be very localized, but affect the person's entire body. I therefore reiterate the need for Yoga therapists to have in-depth knowledge of the anatomo-physiology of the human structure in order to clearly understand the impact of these pathologies. It is vital to know how to treat the injury locally while aiming for overall recovery.

An impairment of the physical sheath can affect a relatively small area of the body. This is considered local. It can also cover a larger area and disrupt an entire system or even, in the worst case scenario, several systems at once. The chronicity of these impairments is of capital importance because the longer they persist, the more the pathology can spread to the surrounding organs and systems.

Most local injuries are acute lesions of the tissues of the locomotor system. Let us examine how Yoga therapists can help individuals with these types of injuries find their way back to health. It is, in fact, a participatory process, as the Yoga therapist often works in conjunction with the attending physician or other health care professional.

Following this, I will present the proposed treatment for chronic impairments affecting all the systems, including degenerative diseases.

Acute Injuries: Sprains and Strains

The first group of acute injuries are part of a set I call sprains and strains. These are lesional states affecting the soft tissues around the joints.

A sprain is a tearing or excessive stretching of a ligament, which renders it unstable or ineffective. A sprain therefore affects a joint's stability.

A strain is an injury of the muscles, usually a tear.

Knowing the major morphological difference between a ligament and a muscle gives us a better understanding of their ability to heal. Ligaments have few blood vessels while muscles are relatively well vascularized. Healing an injury usually involves improving circulation around the affected structure.

However, most ligaments and muscles have an abundance of sensory nerve fibres, making these injuries quite painful.

When faced with an acute sprain or strain, Yoga therapists must inquire about the level of injury in order to adapt their advice.

Grading Sprains and Strains

Sprains and strains have three known degrees of severity:
  - Degree 1: Fibres are stretched beyond normal capacity, micro trauma; little or no tearing (less than 10%); little or no hematoma, pain without swelling.
  - Degree 2: Tear of the tissue (25% to 50%), swelling, limited range of motion, partial or total inability to bear weight, hematoma, very painful.
  - Degree 3: More than 80% tear of the tissue; significant swelling, almost no range of motion, total inability to bear weight, hematoma, sometimes accompanied by avulsion (torn from attachment site) or fracture, very painful.

Yoga Therapy Treatment

Having read these three degrees of severity, it is clear that Yoga therapists need to be aware of their limitations as a front line caregiver.

Unless it is their professional specialty, most Yoga therapists do not have the training required to assess the severity of an acute injury. It is always a good idea to refer clients to professionals capable of conducting the appropriate tests in order to reach an accurate diagnosis. This is especially important when hesitating between the second and third degree of severity. Clinical experience in orthopedic lesions will improve the ability to judge. If in doubt, always refer.

When the client can confirm the degree of severity, the Yoga therapist will be able to act efficiently by adapting the intervention to support healing at each degree of severity.

Yoga therapy for a first degree injury:

The client should rest, stop all âsanas for two or three days to keep the injury from progressing to the second degree and do prânâyâmas and meditation.

There should be a gradual return to physical activity, with the Yoga therapist suggesting âsanas adapted to ensure that each is comfortable. Suggest cocontraction around the affected joints (see the section on cocontraction in the first part of the book). Cocontractions should be gentle to favour better local circulation without putting more pressure on the weakened tissues. The tissue should return to normal within two weeks.

Yoga therapy for a second degree injury:

The client may be taking an analgesic, anti-inflammatory or muscle relaxant. Take great care, as reactions under medication vary considerably from one individual to the next and can mask an aggravation of the condition. It is important to be *cautious* and *observant.*

Weight-bearing and walking are often impossible without support (crutches, a cane, etc.)

It is important to control swelling and pain by applying the classic therapeutic sequence: rest, ice compression and elevation (*RICE*).

RICE is helpful for all sprains and strains of the lower limbs. Elevation is not as important for injuries of the upper limbs.

The tissues need two to three weeks of rest (no âsanas involving the injured area) to improve to the first degree of severity. Again, suggest prânâyâmas and meditation. Rushing the return to activity could cause the injury to progress to the third degree.

Even if recovery appears complete, the injured tissue will often still be sensitive to pressure for a very long time.

Yoga therapy for a third degree injury:
In the third degree, the injured area is completely incapacitated. No weight bearing is possible and the person can only move around with support. He or she may be taking the same medication as someone with a second degree injury and has probably undergone reparative surgery or is wearing a cast. In the latter case, the Yoga therapist should recommend regular cocontractions of the casted limb to minimize the inevitable atrophy caused by non-use.

This phase can sometimes last more than 8 weeks. Yoga therapists can promote healing by proposing anti-inflammatory prânâyâmas (Swara Yoga's Idâ prânâyâma and Shîtalî) and re-energizing prânâyâmas (Matrikâ, Kapâla-bhâti, Bhastrikâ and Breath of Fire).

Continue with the second degree protocol, followed by the first degree protocol.

The affected site may remain sensitive to local pressure for several years.

Other Soft Tissue Injuries
Injuries to the other soft tissues of the locomotor system mainly involve primary inflammation: bursitis, capsulitis, ligamentitis, tendinitis, synovitis, tenosynovitis, fasciitis, etc.

Inflammation is a tissue reaction to an insult (injury, trauma or attack), classically characterized by the presence (obvious or not) of five signs and symptoms. The four cardinal signs of inflammation are *redness, heat, pain* and *swelling.* These classic symptoms have been used in medicine for over two thousand years and are mainly identified by their Latin name: *rubor, calor, dolor, tumor.* The classical authors added a fifth sign: *loss of function* or *functio laesa.*

The signs and symptoms of inflammation are mainly linked to fluid behaviour around the injury. When a tissue is irritated, the blood vessels next to the site will dilate and let white blood cells enter the affected area to help the organism defend itself against the insult that triggered the inflammatory response. This extravasation of the vessels leads to an accumulation of fluid in the tissues (swelling) and more blood (redness and heat). Pain is the result of pressure (from swelling) on the local sensory nerve fibres and by the injured cells releasing irritating chemical substances.

All inflammation involves these signs, but they are not always apparent.

Acute inflammation occurs in the presence of uncommon, repeated movements, sudden, abrupt movements or local impacts. It is sometimes a secondary reaction to chronic systemic impairments.

If there is no tearing or breaks in continuity, acute injuries should heal on their own with rest, ice and light mobilizing exercises. Most acute inflammation will begin to improve in less than five days of complete rest. That being said, complete rest is quite difficult in the day-to-day lives of most active Western adults. Often, in their hurry to heal, injured people are tempted to do too much, too quickly. They exceed the injured tissue's limited capabilities and aggravate their condition. For this reason, it is not rare for inflammation to become chronic.

The Yoga therapy protocol for acute injuries should remain minimal, with insistence on the gentle practice of Hatha Yoga, along with prânâyâmas and meditation.

Chronic Impairments

Chronic impairments are either the progression of unresolved acute states or the classic development of systemic illnesses. Some are identified as degenerative diseases and, depending on their aggressiveness, have more or less permanent side effects and offer limited chances of reversal.

There have been cases in which degenerative diseases have been completely cured. They are still the subject of debate, however, and remain unsolved mysteries for mainstream physiology and medicine science.

In all chronic illnesses, inflammation is the common factor and main cause of chronicity. Chronic illnesses include: cancers, cardiovascular diseases, pulmonary diseases, neurological diseases, autoimmune disorders, fibromyalgia, diabetes, arthritis, arthrosis, Alzheimer's disease, etc.

It is important to remember that all chronic illnesses cause ongoing stress to the structure (connective tissue). This stress gradually transforms the structure into contractile tissue. It therefore has the tendency to retract and lose its suppleness and fluidity.

Remember that inflammation usually develops in an acidic environment that is not well ventilated and usually dehydrated.

Let us examine some of these impairments and how Yoga therapy can help.

Osteoarthritis

Osteoarthritis is the most frequent chronic lesion of the joints. This type of arthritis mainly affects people in middle age or older. It is characterized by the gradual erosion of the cartilage, usually due to uneven rubbing of the surfaces that are in contact. Osteoarthritis is caused by misalignment of the bones or the imperfect congruence of the cartilage in movement and is triggered by poor posture or inadequate weight-bearing (obesity).

In the beginning, there is little pain or inflammation. This intensifies as the cartilage disappears and the bones begin to rub together. As the illness progresses, the bones of the joint swell and deform, resulting in more or less severe ankylosis, stiffness and pain.

The osteoarthritic process is sometimes thought to be the price you pay to live to an old age. I would like to challenge this. Even if it is shown that osteoarthritis begins in adulthood, it is possible to live without any symptoms even at a very advanced age. To keep the structure young and supple, certain behaviours (mainly dietary) and environmental conditions need to be respected.

Rheumatoid Arthritis

Rheumatoid arthritis is another chronic illness, characterized by autoimmune attacks that irritate several joints symmetrically.

It mainly strikes women (70%) between age 30 and 60, but can develop at any age. The weight-bearing joints are the main target, as well as the wrists and hands.

Inflammatory eruptions attack the synovial membrane of the joint, which swells and releases irritating chemicals. The pain is very disabling and clients will often be on anti-inflammatory drugs for an extended period of time. Rheumatoid arthritis progresses intermittently, over several years, even decades. Characteristic deformities can appear in the hands, fingers and all weight-bearing joints. The person must "unlock" the joints in the morning, but this takes increasingly longer as time goes on.

Yoga Therapy for Chronic Impairments

Again, the proposed intervention will be in addition to and in support of care the client is already receiving from other health professionals.

In mainstream medicine, the precise causes of osteoarthritis and arthritis are not well known, but are probably thought to have multiple factors. On the other hand, complementary and alternative medicine believes that the main causes of arthritis, as well as most chronic illnesses, is behavioural and environmental.

If the client has not already done so, the Yoga therapist should suggest dietary changes, especially living foods, due to their alkalinizing and enzymatic properties, as well as encourage adequate hydration.

Regardless of the type of inflammation (local, general, acute or chronic), the client will benefit from energy-giving and anti-inflammatory prânâyâmas (Matrikâ, Idâ, and Shîtalî) focused on exhalation (1 : 0 : 2 : 0), as well as a regular introspective practice.

All arthritis respond well to âsana sequences if they are adapted to the condition. It is the Yoga therapist's duty to select a basic combination of respiration and movement. The focus should be on extending the exhalation time and on the softness of the poses. The poses are meant to maintain range of motion and improve circulation without causing congestion. A gentle practice will ensure this.

This type of daily yogic practice is especially beneficial for elderly people who are most likely to suffer from these illnesses. Not only will they see their symptoms diminish overall, they will feel a subtle strengthening of the muscles around the joints.

Finally, arthritic and rheumatic conditions generally respond very well to heat or, more precisely, a hot environment. Anyone suffering from arthritis will confirm significant relief from pain when vacationing in the tropics and more intense pain in winter or cold conditions. I often suggest that my patients practice some âsanas in hot water (heated pool, whirlpool tub, etc.).

This treatment should be followed for a minimum of 8 weeks. Ideally, of course, the practice should become part of the person's daily life.

The Edge Concept During Self-Treatment

In all chronic conditions (including degenerative diseases), whether general or affecting a specific body part, clients/patients will receive the best results if they follow these simple rules:

- Find the *edge* or *threshold* of pain or discomfort in the affected area. The threshold is the zone just before pain or discomfort. Before this threshold, there is no pain. When the threshold is reached, pain is felt.
- Taking slow, soft breaths, help the mind relax and calm the region by visualizing and feeling the well being of ease, health and joy.
- Slowly, without triggering the pain, cocontract that precise area and hold it in an active stretch, if possible, with Mahâbandha and the Shâmbavî mudrâ.
- Hold for three minutes in Matrikâ prânâyâma, or repeat until you have reached a total of three minutes.

Hypomobility and Hypermobility

Hypomobility

If a client consults for stiffness, the Yoga therapist has several therapeutic options.

Stiffness (also called *hypomobility*) is a state characterized by restricted movement in lower-than-normal range of motion. It can be congenital or developed (through lack of knowledge, lack of training or non-use). Some researchers suspect a genetic predisposition.

Regardless of its cause, stiffness—even congenital—can be improved.

When there is stiffness in a tissue, it is always retracted, as in reaction to stress. The person would therefore benefit from therapy focused on relaxation and improved circulation (massage, bodywork).

Energetic and vigorous prânâyâma sessions centred on exhalation (Matrikâ, Kapâla-Bhâti, Bhastrikâ, Breath of Fire, Pingalâ nâdî prânâyâma, etc.) raise body temperature and cause peripheral vasodilation that increases circulation and warms the tissues. Practicing in a hot environment will help the tissues to loosen. With heat, the retracted tissue tends to stretch more easily.

Hatha Yoga sessions need to be adapted for this condition. The stiff person feels crushed by his/her own tissues. An airy and rather vast practice space must be created so that the person can feel at ease and breathe freely.

The practice must follow the classic tenets, with special emphasis on long Ujjâyî respiration, holding the poses for an extended period (especially in cocontraction, in the passive dynamic phase, see part one) and on the bandhas.

Achieving greater flexibility is a long and arduous process that requires patience. With diligent practice, tangible results should be seen in less than 14 days, although a deep modification of the tissues' suppleness will take a minimum of three months.

Hypermobility

It is quite rare for clients to consult for hypermobility, especially when they seem to be in otherwise good health.

People who are hypermobile can easily perform movements beyond the usual range of motion, without effort or warm-up. The person's connective tissue is generally very loose, especially the ligaments around the joints. This can be due to genetic factors or hormones. (Women are more hypermobile than men because they have a higher concentration of estrogen and a lower concentration of testosterone.)

The main danger for someone who is hypermobile is a lack of stability. Since the ligaments of the joints are not doing their job as stabilizers, they will not be very helpful in protecting from eventual trauma or simply preventing misalignment of the bones at the joints.

Hypermobility is rarely painful. Few people even realize they are hypermobile until an observant person points out the improper alignment of the joints during weight-bearing, imbalance in certain movements or when holding specific postures.

In observing hypermobile people, we can see that, very early on, they develop a tendency to hold postures by *hanging* in their ligaments rather than holding themselves properly with their muscles. Ligaments are meant to stabilize the joint, not hold the person's weight against gravity. Chronically maintaining these postures causes an abnormal loosening of the ligaments and weakening of the muscles due to non-use.

Changing this state is even more difficult than with hypomobility. One of the main reasons is that hypermobile people often have weak muscles. Slow re-education therefore requires overall strengthening as well as precise strengthening around certain key joints.

Re-education is a slow process because strengthening a muscle takes time, but also because the tendency to return to hypermobility always remains, patiently waiting. The person must wage an ongoing battle against hypermobility.

In Hatha Yoga, the focus should be on strengthening, by holding prolonged cocontractions while breathing smoothly. And, most importantly, always maintaining a healthy alignment.

Other Chronic Diseases

All medical pathology texts give a logical, ordered classification of all known diseases. Among the most classic chronic diseases, I present a list of the main conditions below. It is my belief that

Yoga therapy, not being limited to the physical sheath, is an effective tool for improving all of these conditions.

Neurological diseases
Hemiplegia, paresis, dementia
Cardiovascular diseases
Stroke, angina, infarction, heart failure
Respiratory disease
Chronic obstructive pulmonary disease (COPD), (bronchitis, emphysema), asthma
Digestive diseases
Liver and gallbladder diseases (gallstones, jaundice, hepatitis, cirrhosis)
Gastric diseases (ulcers, pyrosis)
Pancreatic diseases (type I diabetes, pancreatitis)
Intestinal diseases (Crohn's disease, ulcerative colitis, irritable bowel)
Genito-urinary diseases
Sexually transmitted diseases (STD), vaginitis, prostatitis, cystitis, nephritis
Degenerative diseases
Cancer, multiple sclerosis
Infectious and viral diseases
Add unwanted bacteria or inopportune viruses to this list for another set of illnesses in each of these systems.
Autoimmune, behavioural and social diseases:
Depression, bipolar disorder and other neuroses, chronic fatigue, fibromyalgia, obesity, type II diabetes, lupus

These conditions are also classified by medical specialty: orthopedics, neurology, gynecology, urology, pediatrics, gerontology, etc.

It is not the aim of this book to explain all the ways that Yoga therapy can help alleviate these conditions. For several of the diseases above, there are reliable studies that demonstrate the benefits and validity of this therapeutic approach beyond all doubt.

Know that all chronic diseases, whether developed or innate, genetic or congenital, can be improved through a healthy diet, appropriate hydration, prânâyâmas and the regular practice of this protocol, within the limits of the condition. With Yoga, the condition can always be transcended.

Building a Yoga Therapy Program

Applying the protocol to the three bodies should be highlighted by the Yoga therapist's ease in helping transform the physical structure. This requires knowledge of how to build the correct sequence of âsanas and prânâyâmas.

Normally, this science should be taught in all professional training courses for Hatha Yoga teachers. However, not all Yoga therapists are duly accredited Hatha Yoga teachers. I therefore suggest a simple and effective means of putting together a physical Yoga therapy program.

Prerequisites

The initial rule is to respect the limitations or contraindications of the client's condition.

The chosen âsanas and prânâyâmas should not increase inflammation, blood pressure, hypotension or extend discomfort or pain. Yoga therapists must know how to adapt the exercises to these conditions.

Women

It is especially important to check if your client is pregnant or experiencing a painful disruption of her monthly cycle. If that is the case, the sequence of âsanas and prânâyâmas should be modified or adapted accordingly. Contrary to what is sometimes taught in certain schools, it is absolutely not contraindicated to practice Yoga during the menstrual cycle. However, it is important to adapt the practice to the woman's condition.

In the last 30 years, in the West, Hatha Yoga has been mainly practiced by women. Yoginîs have therefore developed vast expertise on the subject. I invite you to consult their yogic knowledge and the wisdom they have amassed, more specifically on the relationship between Yoga's sâdhana and women's health.

Therapeutic Program

All âsana or prânâyâma sequences should begin with a conditioning of the physical sheath. The patient centres him or herself as best possible (dhâranâ, Sukha and Ujjâyî prânâyâmas, drishti) and settles comfortably, ideally with a straight spine.

If the patient has a serious condition, is unable to come to you or is bedridden, it maybe be a good idea to begin a return to health via Sukha or Matrikâ prânâyâma, followed by dhyâna, then a visualization of the ideal therapeutic sequence. This is a wonderful mental exercise. Like a high-level athlete visualizing each move before a competition, the patient performs each âsana mentally with all the ease and fluidity of unparalleled fitness and power. This strategy is very effective at encouraging the subconscious to actively support the return to health.

Prânâyâmas

I cannot overstress the remarkable therapeutic potential of prânâyâmas. Prânâyâma Yoga is probably the most powerful tool in Yoga's therapeutic arsenal and is, in my opinion, the most powerful yogic tool, period.

I usually recommend that all therapeutic sessions begin with prânâyâmas. This will calm the mind and prepare the body for the âsanas that follow.

The nomenclature in section *Preparing to Practice Prânâyâmas* of this document is sufficiently detailed to form a solid initial reference for building a therapeutic prânâyâma session. Some points to remember:

- If the person is agitated, begin the therapeutic program with prânâyâmas that will quiet the mind and body. Have the person perform Sukha, Matrikâ and Ujjâyî for a few minutes or 15 to 25 cycles.
- Always stop the practice if there is discomfort, nausea or pain. Have the person lie on their back in savâsana and rest until the symptoms disappear.
- Apply the appropriate practice to bring about the desired effect in the body or specific system.
- Always combine the practice with visualization in the body or mind.
- End the session with several minutes of an integration dhyâna.
- Then, proceed with the âsanas.

Sun Salutation

Often, at this initial stage, performing the series of poses that make up the *Sun Salutation* can be very helpful, as this exercise covers every area of the body, creating an ideal lead into the more specific therapeutic âsanas. A dozen repetitions should prepare the body (warming it and loosening it up). Sometimes, depending on the client's condition, the Sun Salutation alone (or even a simplified version) will be the therapeutic self-treatment exercise of choice.

Sequence of Âsanas

Following are some important points to teach patients in building all therapeutic âsana sequences:

- Most poses begin in a standing position and end in a standing position.

- Insist that each pose be performed in stages. It is important to master each step before moving on to the next, more advanced step. Each step of an âsana is an âsana in itself.
- Never force an âsana. Always stay within the threshold of comfort.
- Work in pose and counter-pose pairs.
- Relax between each âsana, most often in savâsana, for the same amount of time the pose was held.
- Be aware of the three phases in practicing each âsana.
- Make sure that the structures are properly aligned.
- The âsana*s* should be performed slowly in order to observe the inner sensations (body and mind) that arise from this practice.
- Be sure to keep respiration fluid at all times. Do not hold your breath.

To Touch or not to Touch

Most Yoga therapists have considerable Yoga experience. Many have taken courses to become Yoga teachers; others are already bodywork practitioners or health care professionals. Yoga therapists need to decide whether to treat their clients through verbal advice alone or if it would be relevant to intervene with approaches that involve physical contact.

The answer? It depends.

For Yoga therapists who are already health care professionals and are comfortable treating physically, this is not a problem. They can, with the client's consent, incorporate relevant physical interventions that support the Yoga therapy.

Yoga therapists who do not have any prior physical treatment experience should keep their interventions within the realm of suggestions, guided demonstrations and contact-free energetic healing techniques

Some cultural differences should also be taken into account. For example, in some countries or in some areas of the world, it is not appropriate for a man to treat a woman and vice versa.

Adjustments

A good Hatha Yoga teacher must be able to correct a pose or propose adaptations in accordance with the client's specific needs. I believe that good Yoga therapists working with âsanas should also have this skill.

In teaching Yoga therapists, there could be a knowledge base involving the pose adjustments required to perform âsanas and as well as other suggested therapeutic poses. The art of making adjustments lies in suggesting corrections without guiding too rigidly. It is an art that is transmittable and teachable.

CHAPTER VI

PROPHYLAXIS

Prevention and Maintenance

Everyone who applies this therapeutic protocol will immediately experience improvements in balance and health. However, the best therapy is still prevention. Practicing Yoga when already in good health produces remarkable vital power.

Nutrition and Hydration

Health is always the result of several factors. In the first part of the book, we saw that the main factors in vitality's physical expression are a healthy diet and good hydration in a harmonious environment.

Fasting

On the subject of food, several Yoga schools propose that their followers fast one day per week. This method, sometimes called *intermittent fasting*, appears to be one of the keys to longevity and health.

Fasting is an extremely healthy preventive or curative practice for most people. It promotes internal cleansing, renews the organism's vital energy and gives the digestive system a rest.

Recent research has demonstrated that intermittent fasting is a very efficient means of reducing inflammation, stimulating the immune system and supporting tissue regeneration.

Mauna

Some schools also suggest a very powerful practice for renewing one's vital energy: *mauna*, silence!

Observing one day of silence per week (perhaps on the same day as the fast) has an undeniable positive effect on prânic flow. It is incredible how much energy is devoted to speech.

Some practitioners voluntarily practice asceticism (tapas), choosing to stop speaking altogether (vow of silence).

Practicing mauna one day per week (or more, if you like), truly recharges the batteries!

Personal Hygiene and Shatkarmas

Yoga therapists should be familiar with and able to teach Yoga's physical hygiene techniques. These are very ancient purification practices described in the classical Hatha Yoga texts (*Hatha Yoga Pradipika, Gheranda Samhitâ, Shiva Samhitâ*, etc.). There are six fundamental practices: *Shatkarmas* (some authors call these *Shatkriyâs*): *Trâtaka* (concentration and eye purification through intense observation without blinking), *Neti* (cleaning of the nasal passages and sinuses), *Dhauti* (cleaning of the mouth, ears and digestive tract), Kapâla-bhâti (purification and revitalization of the upper airways), *Nauli* (abdominal massage) and *Basti* (colorectal cleanse).

These activities have been a part of yogis' sâdhanas since time immemorial, likely with good reason.

In this section, I will be explaining a few of these Shatkarmas as well as other complementary practices. In addition to maintaining the physical sheath, these actions contribute to the subtle purification of the being and can instill a greater understanding of the famous saying: *a sound mind in a sound body*.

Some of these activities are somewhat controversial. If you are ill, it may be risky to try them on your own. It is preferable to seek the guidance of an experienced yogi or yoginî. If you are not sure if these practices are appropriate for your condition, I advise you to consult a doctor or health care professional who is familiar with these methods. Remember that these techniques were developed at a time in which hygienic knowledge and conditions were quite different. Use your judgement.

My patients are always pleasantly surprised by these alternative methods of hygiene, finding them to be simple, natural solutions to several physical dysfunctions, most without side effects.

The âsanas presented here are not the only ones that exist. Each âsana has a whole set of variations. I am only suggesting a few. Know that the therapeutic practice of âsanas almost always leads back to basic poses, those usually considered to be beginner poses in the various Yoga schools and ashrams.

Each therapeutic situation is unique. Yoga therapists need to know how to adapt the poses and techniques to the patient's condition.

Here is a brief overview of the actions, from head to foot.

The Skin and its Appendages

The epidermis is made up of several layers of epithelial cells. The deepest layer is the *stratum germinativum*. It is made up of cells that constantly reproduce. The new cells push the old ones to the surface layers, up to the periphery (surface of the skin). This continual movement is the body's way of renewing our external sheath. So much so, that the skin you feel today did not exist last month!

The cells of the external layer of the epidermis are not irrigated by blood circulation. This layer is essentially made up of dead cells. They are continuously pushed up by the new cells and pile on top of each other, in a design that looks a little like fish scales or roofing tiles. The dead epithelial cells are eventually eliminated from the body through the physiological process of desquamation, the skin's self-cleaning mechanism. When you wash your body, you are stimulating this natural activity.

Even though the cells of the epidermis are dead, they still form an essential protective barrier against external attacks. Over 10 million bacteria cover every square centimetre of skin! These live in symbiosis with the outside of the body and contribute to its protection and overall health. However, they become dangerous pathogens the moment they pass through the epidermis. This can occur when there is an accidental or trauma-induced tear, causing infections that can be lethal.

Between the epidermis' rows of stacked cells, there are several openings that reach down into the deep layers: the pores and hair follicles.

Pores are the excretory channels of the sweat glands located deep in the dermis (the layer of connective tissue under the epidermis). One of the body's excretion and cooling methods, perspiration occurs when these glands secrete sweat that is carried to the surface of the skin. For these reasons, perspiration should never be blocked or prevented from occurring. On the contrary, it should be encouraged!

Hair follicles are also anchored deep in the dermis, but their channel contains a hair. Hair covers the vast majority of the body's surface. Each follicle has a sebaceous gland that produces an oily secretion on the surface of the skin. This sebum lubricates the hair and covers the entire surface of the epidermis. It ensures that the skin is well hydrated by locking in humidity. Sebum forms a protective, antibacterial film and is an effective barrier against the constant bombardment of dust, viruses and fungus spores. This natural oil also helps hair and skin better resist UV damage and outdoor pollutants.

Soaps and Body Oils

Since the beginning of time, humanity has created quite an economy around bathing and cleaning the skin. Men and women of all cultures have developed numerous ways to favour or even improve the art of getting cleaner.

Soap, whether solid or liquid, is a surface active substance made up of molecules that can bind to fat and water. Dead cells and accumulated dirt and grease are diluted and trapped by the soap and carried away during rinsing.

The problem with soap is that it often leaves the skin dry and fragile until it has rebuilt its sebaceous film (this takes several hours). Moreover, today's soaps, sold on the lucrative cosmetics market, are very powerful, often harmful cleansers. They contain detergents whose molecules are sometimes even carcinogenic.

Soaps used to be plant-based and were rarely a threat to health. That is no longer the case. Many blends are concocted in factory laboratories and are far removed from natural substances. The cleaning products contain petroleum-based molecules, emulsifiers and irritating foaming agents, such as sodium lauryl sulfate, parabens and other preservatives. These products have multiple toxic properties that are increasingly criticized. Several have been blamed for contributing to the rise in cancers in our so-called "advanced" societies.

Recipe

Yogis and yoginîs have used a variety of very healthy body cleansing and purification methods for thousands of years, incorporating these practices in their daily sâdhana.

One of these practices (still popular in southern India today) involves lubricating the body with virgin sesame oil and cleaning the skin with mung bean powder (or mungo). The sesame oil blends perfectly with the sebum, deeply penetrating the skin and softening the tissues. It is also an important source of vitamin E, an antioxidant that is said to have antineoplastic properties.

The mung beans are ground into a powder in a coffee grinder. Before, this was done manually, with a mortar and pestle. A little coarser than flour, the powder is a pale green with yellow flecks (granules). It foams slightly in water and its granules act as the abrasive needed for cleaning. The procedure is as follows:

> – Moisten the skin with a little hot water.
> – Generously lubricate the epidermis with virgin sesame oil (preferably organic). Massage to work into the skin.
> – Sprinkle the mung bean powder on the skin. Rub briskly.
> – Rinse with hot water.

The skin is completely regenerated after this treatment! Clean, soft and satiny, it literally breathes through all its pores, providing a very pleasant sensation of natural freshness.

Some dermatosis are soothed by this practice.

In Asian countries, especially India, sesame oil is also used on the head to tone and soften the scalp and make the hair strong and shiny. However, from experience, I do not recommend using mung bean powder in your hair because it will form a mass of particles that are extremely difficult to dislodge during rinsing. Other natural foaming agents for the hair can be found in stores that sell natural health products.

Be advised that the mix of powder and oil can block bath or shower drains.

Massages

The physical sheath's highly connective tissue structure requires constant maintenance. In order to be supple, the collagen fibres and all the soft tissues need to be warmed through mobilization.

Mobilizing the soft tissues encourages the circulation of blood, lymph and all other fluids. This results in enhanced nutrition and an increase in energetic and prânic fluidity. In short, better overall health.

All of the body's tissues are stimulated in a healthy Hatha Yoga practice. However, other physical practices can do the same: dance and martial arts, for example, as well as all bodywork.

Massage is one of these many approaches. It encourages fluid and energetic irrigation and detoxification.

Massage can take place between a giver and a receiver, with all the psycho-energetic exchanges this could involve. But it can also be a self-practice, in which the various body parts are relaxed and stimulated.

In any case, massage is a wonderful alternate modality that promotes tension releases.

Yoga of the Eyes

The eyes are the organs that send the most sensory information to the brain. These impulses are spread throughout the sensory cortex of each hemisphere, indicating the important role visual information plays in cerebral sensory integration. Not only is it possible for the eyes to perceive light, but the visual pathways, which are connected to the different sensory cortical areas, enable the interpretation of these light waves, giving them meaning and significance. The images built in the visual cortex and other sensory cortical areas (visual associations) form a vast warehouse in which mental information is stored.

This store of images gives us the ability to visualize internally. Our mind is able to create new images or use memorized images in new contexts or mental environments (reflection, visualization, etc.).

Several yogic activities involve the eyes. In fact, each of these activities could also be used to develop eye health and vision. Following are some yogic practices that call on the eyes and benefit visual function.

Trâtaka

This activity is essentially a dhâranâ (concentration) exercise. It is also one of the six Shatkarmas. Several variations exist. This is probably the most well known:

In a shaded, quiet room, sit in a stable and comfortable position, with a straight spine. Place a candle at heart height (lower than eye height), at a distance no further than your outstretched arm (about 40 cm). Distance will vary from person to person, but should remain comfortable.

Method
- Close your eyes. Apply the Shâmbavî mudrâ for several slow respiratory cycles. Gently open your eyes.
- Stare at the heart of the flame, at the small space between the lit wick and the dancing flame.
- Do not furrow your eyebrows or blink. Breathe calmly.
- Without taking your eyes from the centre of the flame, concentrate around the flame, on its flickering halo, with your peripheral vision. Your gaze should be immobile and focused on the centre of the flame while your attention is on its periphery.
- Hold this gaze for 15 seconds. Then close your eyes and observe, without forcing, the after-image of the flame on your retina, until it disappears.
- Repeat three times.

In the beginning, tears may blur your vision. But with practice, you will be able to lengthen the session up to one minute without effort and be able to observe the inner image for over four minutes.

Any time you feel an unpleasant burning sensation, simply close your eyes, observe the after-image and resume when the discomfort disappears.

This exercise is known to help develop creative visualization. You will be gradually increasing your perception through the third eye, Âjnâ chakra. Often, this phase of trâtaka will lead you into dhyâna.

You will develop not only remarkable concentration abilities, but also better visual acuity.

Suryadhyâna

Another practice stemming from trâtaka is *Suryadhyâna*. In this inspiring activity, the sun replaces the flame!

This is an extremely powerful practice, with infinite potential.

Method
- This activity is practiced outdoors, until one hour after sunrise or within one hour of sunset (dawn and dusk).
- In a calm, natural environment (ideally), sit or stand in a comfortable position, facing the sun. Clouds must not obscure the sun during this exercise.
- Observe the sun in the same way as the flame in trâtaka. Start with 10 seconds.
- You can blink, but do not furrow your eyebrows. Look at the sun without moving or straining. The minute you feel discomfort, close your eyes and then gently resume looking at the sun.
- Add ten seconds every day you practice.
- This practice will help you quickly enter a meditative trance.
- At the end of the session, close your eyes and observe the afterimage of the sun between your eyebrows. Gaze inwardly for as long as you like.

Some authors, like Hira Ratam Manek, claim that all illnesses will be cured if you work up to holding the gaze without strain for 45 minutes! This would take about nine months of uninterrupted sun to achieve, if you add ten seconds every day.

Cloudy or rainy days are days off!

There are other exercises that involve the eyes. The various forms of *drishti* (fixed gaze) are very useful, not only for improved concentration, but also to keep the eye muscles strong and supple. Shâmbavî mudrâ is the ultimate drishti, in which the gaze is turned inward, between the eyebrows, at the Âjnâ cakra. This mudrâ accompanies most Yoga practices that require the eyes to be closed.

In addition to these yogic exercises, the *Bates eye re-education method* is very well known. It includes several exercises meant to naturally restore eye health. (The Bates method has inspired many books and courses).

Nasal Irrigation

Neti is the yogic practice of cleaning the nasal passages and sinus cavities.

The bones of the face have cavities in their structure. These cavities, the bony sinuses, communicate with the outside through small bony canals that end in the nasal cavity. They are the *frontal, ethmoid, sphenoid* and two *maxilla* bones.

The meatus (opening) of these canaliculi in the nose is protected by three bony folds of the nasal's cavity lateral wall, the *nasal septum.* The two upper folds are the folds of the ethmoid bone and are called the *superior nasal* and *middle nasal concha.* It is these bony excrescences that house the meatus of the sinus canaliculi. The lower fold is a separate bone and makes up the inferior nasal concha. It is not connected to the sinuses, but protects the meatus of the lacrymal canal.

All of the walls in the nasal cavity, including the median septum and concha, from inside the nostrils to the nasopharynx at the back of the mouth, are covered in a very humid mucosa. This mucosa also covers the walls of all the sinuses, the Eustachian tubes and middle ear. It spread as far as the larynx and overflows down the throat to cover the inside of the respiratory airways and digestive tract.

This mucosa is humid because it secretes relatively aqueous mucus with multiple functional properties.

When the mucosa is irritated, the mucus cells react by secreting even more mucus to protect the wall. This causes a thickening of the layer of mucus for better mechanical protection. If this is not sufficient, the wall becomes inflamed and may be more vulnerable to the incessant viral and bacterial attacks and succumb to infection.

Unfortunately, mucus hypersecretion has its inconveniences. Aspirated dust and other impurities (bacteria, spores, etc.) are more likely to stick to it. Also, the thickening obstructs airflow, which can sometimes cause choking. This state can be as harmful for health as the initial reaction was trying to prevent.

Blowing one's nose is the classic way to clear mucous obstruction. However, even this healthy action becomes an irritant if you have a cold or flu. Colds and flu are nothing more than a viral inflammation of the nasopharyngeal mucosa, which can develop into generalized irritation of the nasopharynx (pharyngitis), to the larynx (laryngitis), sinuses (sinusitis) and even the bronchi (bronchitis). If the infection is more severe, a cold can develop into the flu and spread all the way to the lungs (pneumonia). The infectious agents in these cases are viruses and bacteria.

The conventional medical means of ridding the body of bacterial infection is pharmacodynamics, more specifically, antibiotics.

There are few medical treatments against viruses. The usual method is based on vaccination theories (widely contested by defenders of alternative approaches). The complementary and alternative medicines favour the use of natural medicinal agents that boost the immune system.

Yogis and yoginîs have developed an original method of protecting the nasopharyngeal mucosa: sinus irrigation or Jalaneti.

Jalaneti

For this practice, you will need a container that looks like a small tea pot and can contain one or two cups of liquid. Most Yoga ashrams and Yoga stores sell specific containers (made of plastic, ceramic or metal) called *lotas* (Neti pots).

You will also need natural, unrefined sea salt (fleur de sel, Himalayan salt, etc.).

Method
— Fill the container with body temperature water (around 37-40° C).
— Add one teaspoon of salt per 2 cups of water.
— Let's randomly start on the left side.
— While standing upright, place the spout of the lota (or small tea-pot) in the opening of the left nostril (so that the water can flow freely into the left nostril when you bend over). The spout should completely fill the nasal opening.
— Bend over the sink and turn your head to the left so that the right nostril is at the bottom (closest to the sink).
— To keep from swallowing the water and choking during this process, breathe gently and evenly through the mouth.
— The warm, salty water will flow into your left nostril, through the nasal passages and sinuses on the left side, around the back wall of the nasal septum and emerge from the right nostril in a continuous flow into the sink.
— When one cup has been emptied on the left side, remove the lota; turn your head toward the sink and blow a dozen times with both nostrils. This will expel the water accumulated in the passages. Use another cup of water on the right side.
— When the lota is empty, turn your head to the sink and blow a dozen times with both nostrils, either over the sink or into a tissue.
— Stand upright and dry your nose again by blocking one nostril and blowing with the other a dozen times. Repeat on the other side. Blow as often as you need to dry your nose.

If you do this regularly, it will eliminate the likelihood of developing colds and flu. Jalaneti also improves asthmatic conditions and most otorhinolaryngological dysfunctions. Like all of the previous techniques, this is a preventive practice. However, if you are already weakened by an illness affecting this area, Jalaneti will be very helpful in supporting healing.

Be sure to use good quality water. In case of need or when in doubt, filter it, boil it or use distilled water. Do not share your lota with another person, as this could cause infection by contagion. Always carefully clean your pot after every use.

Dhauti

Dhauti is one of the six Shatkarmas. There are several variations, all for mouth, ear and digestive tract hygiene. Here are a few.

Danta Dhauti

This practice cleans the external auditory canal and the mouth.

Rinsing with warm water and using the nail of the little finger efficiently clean the auditory canals. Be careful not to penetrate too deeply. Do not use a swab or anything metallic to clean the ear as this could damage the walls of the auditory canal or even pierce the eardrum.

If there is too much cerumen (wax), or if it is dry and difficult to dislodge, add a few drops of mullein or sesame oil to soften. Let penetrate for a few minutes. If the cerumen resists, use a jet of water. This last action should be supervised by an experienced person or performed by a doctor.

The mouth is one of the body's main entryways. Billions of bacteria team within it as well as dust and all kinds of microscopic debris, without forgetting the solid food, liquid food and secretions you are continually swallowing. It is therefore important to keep this area clean and in optimal condition.

Brushing daily and using dental floss maintain the health of the teeth and gums. Some yogis and yoginîs use plant roots or stems instead of commercial brushes. In addition to their mechanical benefits, these stems (often from the neem tree) offer significant protection when chewed due to the various astringent and medicinal effects the tannins in the sap have on the gums and roots of the teeth.

Beware of conventional toothpastes, which often contain toxic substances such as fluorinated molecules and sodium lauryl sulfate foaming agents. There are natural alternatives.

One part of the mouth that is often neglected in oral hygiene is the tongue! This highly sensitive, muscular organ (over 14 muscles), which plays a key role in mastication and speech, is dotted with mucus glands, salivary glands, various tactile sensory receptors and taste buds.

The tongue is covered in a layer of mucus and saliva. This film is very high in bacteria and debris of all kinds, which can even affect its sensory functions. It can also become an important contributor to *halitosis* (bad breath) or *cacostomia* (bad odour emitted from the mouth).

Yogis and yoginîs often combine their Jalaneti practice with their *Jîva Dhauti or Jîvamûlasodhanam* (*tongue cleaning*) practice.

Jîva Dhauti (Jîvamûlasodhanam)

This practice involves cleaning the tongue with a scraper. You can use a small metal instrument designed specifically for this purpose or simply an upside-down coffee spoon.

Rinse the scraper before and after each use. You can do this practice in the shower or over the sink.

Method
— Place the scraper as far back on the tongue as possible, being careful not to trigger the gag reflex.
— Scrape the surface of the tongue from back to front, pulling the mucus out of the mouth. Spit.
— You can then spread a bit of clarified butter (ghee) or sesame oil on the tongue. Let penetrate.
— Scrape the surplus off and spit again.

The clarified butter or oil will keep the surface of the tongue from becoming irritated.

You will feel an immediate sensation of freshness on the tongue, with a guaranteed improvement of all the sensitive lingual functions.

This technique is also performed with the fingers, using the nails and fingertips as the scraper. Be sure to carefully wash your hands beforehand.

Other Practices for the Head and Neck

Practicing *Simhâsana* (roaring lion pose) on a regular basis tones and softens the tongue, the temporomandibular joints and the eye muscles and provides appreciable mechanical stimulation of the vagus nerve and its parasympathetic functions.

It is said to benefit the voice by toning the laryngeal muscles and softening the vocal cords. It is also thought to have a regulating effect on the thyroid gland. Brâhmarî prânâyâma produces similar effects.

Yoga offers another extremely powerful laryngeal practice. It is part of Nada Yoga, the Yoga of the primordial sound. Begin by inhaling. As you exhale, chant the sound *"Om"* out loud. Follow the vibratory resonance of the sound by closing the mouth and drawing out the "m", first in the skull, then along the spinal column and, as applicable, the entire structure of the body. Hold the sound vibration for as long as is comfortable. Then hold the silent reverberation of this vibration within, as an opening to dhyâna.

You can also use the *"Aum"* sound. As you exhale, detach the sound of each letter, then, by closing the mouth, draw out the "m". The "a" becomes an "oo", then, as the mouth closes, an "m". According to yogic literature, the Yoga of Sound provides all the benefits of the other types of Yoga.

All Hatha Yoga poses that involve cervical flexion and extension will have a powerful physical effect on the flexibility, strengthening, circulation and innervation of this area. Poses include: *Viparîta-karanî (inverted pose), Sarvangâsana (shoulderstand), Halâsana (plough), Sasangâsana (rabbit), Matsyendrâsana (fish pose), Sîrsâsana (head stand)* and all their variations. Some of these poses are combined with Jalandhara bandha.

Thoracic Hygiene

The thoracic region is bordered, from the front to the back, by the rib cage, with its twelve pairs of ribs, sternum and clavicles (which are also part of the upper limbs), the twelve thoracic vertebra and the shoulder blades (also part of the upper limbs). This region is delimited at the top by the thoracic inlet and at the bottom by the diaphragm muscle. The rib cage houses the heart, thymus gland, lungs, esophagus and trachea. The aorta and all of its arterial branches emerge from the heart and make their way toward the head, upper limbs and diaphragm. All of the arteries of the thorax do the same. The heart also sends important arteries directly to each lung (the pulmonary arteries). The superior and inferior vena cava and the pulmonary veins drain into the heart. The azygos veins, on either side of the thoracic spinal column, as well as the main lymph ducts, also flow toward the heart.

The thoracic spinal cord is the control centre for the sympathetic innervation of the autonomic nervous system and, from there, forms two chains of ganglia that travel longitudinally on either side of the column in the thorax and abdomen. This autonomic structure is the source of all sympathetic innervation and is distributed in every part of the body.

The multiple branches of the vagus nerve (tenth cranial nerve) are responsible for parasympathetic innervation in the thoracic cavity. From the skull, the vagus nerve descends into the neck on each side and enters the thorax through the thoracic inlet. It continues down in a spiral around the esophagus, all the way to the diaphragm.

The peripheral thoracic nerves innervate the muscle structure of the torso and a good part of the abdomen. They are also responsible for skin sensitivity in this area.

All Hatha Yoga poses involving the thoracic region will soften the vertebral and costal joints and their muscles. In so doing, they contribute to the suppleness and proper functioning of the thoracic spine and nerves.

Thanks to simple gravity, inverted poses stimulate circulation and better drainage and also cause the viscera to move slightly, helping tone these according to their new positions. Thoracic twists improve organ suppleness and function, increasing their range of movement and deformation in a very healthy internal self-massage.

Hrid Dhauti

There is an advanced esophageal and gastric tract cleansing practice called *Hrid Dhauti*. Hrid means heart, referring to the heart area or the heart cakra. The practice is a series of purifying actions focused on eliminating mucus and waste from the walls of the esophagus and stomach.

The texts claim that, in addition to cleaning and toning the esophageal and gastric tract, Hrid Dhauti stimulates and releases tension in the mediastinum, bronchi, bronchioles, lungs, heart and thoracic vessels, from the top of the back and shoulders.

The first of these actions is *Danda Dhauti*. It involves inserting a small, flexible tube into the mouth, past the laryngeal glottis in the esophagus, all the way to the stomach. Remove slowly. The tube's friction cleans and tones the esophageal walls. Another variation is to insert a small catheter in the same way after drinking several glasses of salt water. With a few Nauli-type abdominal movements, the water is expelled through the catheter. *To be practiced only under the supervision of an experienced yogi or yoginî.*

The next technique is called *Vamana Dhauti*. This practice does not require any tools. All of its variations involve stimulating the gag reflex as an elimination practice. Vamana means to vomit. On an empty stomach, drink three or four glasses of salty water. After mixing abdominally with Uddîyânabandha and Nauli kriyâ, induce vomiting by stimulating the back of the throat. The water is completely ejected from the stomach, along with excess mucus and accumulated waste.

This practice's known effects extend beyond the stomach. As in Danda Dhauti, it relaxes the gastric, pulmonary and cardiac functions as well as the entire otorhinolaryngology. It also balances the five prânas, especially prânavâyu, the ascending prâna. Vamana Dhauti acts as an antidepressant and combats mental heaviness.

A variation of Vamana Dhauti is *Vhyâghra kriyâ*. Vhyâghra means tiger. Just as tigers sometimes regurgitate their food or the hair they have ingested, the technique involves inducing vomiting of the gastric contents approximately 90 minutes to 3 hours after eating. The benefits are the same. *To be practiced only under he supervision of an experienced yogi or yoginî.*

Vastra Dhauti

This practice is probably one of the most impressive of the Shatkarmas. Some authors relate it to variations of Hrid Dhauti. Vastra Dhauti involves swallowing a long ribbon or band made of linen or finely woven cotton, about four fingers wide and about 3 to 5 metres long (10 to 16 feet). The fabric is moistened with salt water then gradually ingested by taking small swallows with sips of salty water, until only 30 cm (1 foot) or so remains outside of the mouth. The ribbon fills the entire stomach and esophagus.

Stand straight and perform the Nauli kriyâ for about five minutes. Then gently pull out the ribbon. It will be more or less covered in mucus and debris from gastric digestion.

Vastra Dhauti is often practiced after Vamana Dhauti. In addition to its obvious cleansing effects on the stomach and esophagus, Vastra Dhauti produces the same benefits attributed to Hrid Dhauti. It is an effective therapy for asthma, pulmonary afflictions and dysphonia (voice problems).

The Dhauti techniques, particularly those that use vomiting as a means of purification, are not looked upon favourably in our modern societies. These ancestral yogic methods may be misunderstood and used incorrectly, such as in the disorders of bulimia and anorexia. Once again, if you are interested in exploring these practices, it is advisable to do so under the supervision of an experienced yogi or yoginî.

Asanas of the Thoracic Region

Some âsanas have significant effects on the thoracic area: *Sîrsâsana*, Sarvangâsana, Halâsana, Viparîta-karanî, *Bhujangâsana (serpent), Dhanurâsana (bow), Cakrâsana (wheel), Trikonâsana (triangle), Marîcyâsana (seated twist, Sage Marîcy's pose)* and all their variations.

In fact, all the types of prânâyâmas make an important contribution to thoracic health. Each of the âsanas I mentioned is synchronized with a form of prânâyâma, some of which are excellent at clearing the lungs. According to several authors of classic texts, Kapâla-bhâti, one of the six Shatkarmas, is a kriyâ for the purification of the airway.

Abdominal and Pelvic Hygiene

The abdominopelvic cavity starts with the diaphragm muscle at the top and ends in the pelvic floor at the bottom, essentially made up of the *Levator ani* muscle. The lumbar spinal column and the sacrum and coccyx form the posterior bone wall of this area while the abdominal muscles make up the bulk of the anterior abdominopelvic wall. The two hipbones complete the pelvic region's bony pelvis.

Under the diaphragm, the abdomen holds several organs, which mainly have digestive functions: liver, gallbladder, stomach, pancreas, small intestine and colon. The spleen is located slightly under the stomach, at the left lateral extremity of the abdomen and is chiefly related to blood regeneration.

The kidneys are located on either side of the lumbar column, with the adrenal glands sitting on top. The ureters emerge from the kidneys and extend down, passing in front of the psoas muscles and into the pelvic cavity, where they descend to the back of the bladder. The bladder is right behind the pubis, on the floor of the anterior perineum. Its urethra passes through the perineum and outside the body.

In men, the rectal ampulla and anus are located behind the bladder, in front of the sacrum and coccyx. Under the bladder, on the urethra's path, we find the prostate, whose posterosuperior wall also receives the tubes of the two seminal vesicles and the two vas deferens of the testicular spermatic cords.

In women, the uterus is located between the bladder and rectum. The Fallopian tubes and ovaries are located on either side and slightly to the back of the uterus. The cervix forms most of the vaginal ceiling, whose lower opening (vestibule) is located directly behind the urinary meatus. In adult women, the uterine fundus is usually tipped forward over the bladder, about two finger widths above the pubic crest.

All of these organs, including the reproductive organs, are innervated by sensorimotor nerves emerging from the lumbosacral spine and by the autonomic nervous system, originating from the ganglia of the solar plexus and the hypogastric nerve plexi.

Arterial blood flow is provided by the branches of the abdominal aorta, which crosses the diaphragm at the 12th thoracic vertebra (T_{12}). Blood flow in the iliac veins, arriving from the lower limbs, enters the pelvis at the groin. The iliac veins meet in front of the 5th lumbar vertebra (L_5) and become the inferior vena cava, which drains all the abdominopelvic organs by rising across the diaphragm toward the heart. Lymphatic drainage is provided by all the lymph nodes and abdominopelvic collector vessels that drain at the cisterna chyli, behind the aorta, in front of T_{12}.

Agni Sara Kriyâ (Vahni Sara Dhauti)

There are several variations of this practice, all aiming to quickly build abdominal fire or Samâna prâna. I mentioned one of these, Breath of Fire, in the different types of prânâyâmas. Breath of Fire is, in fact, a variation of Agni sara kriyâ. The intense heat purifies the gastrointestinal tract. Mastering Uddîyânabandha is very helpful in accomplishing this.

The technique is relatively simple:

> — While standing, place your hands on your thighs and lean slightly forward.
> — Forcefully expel the air from your lungs and, in kumbhaka (holding the breath) pull the abdominal wall back and up, through negative pressure, producing Uddîyânabandha.
> — Repeat this several times without breathing, until discomfort forces you to stop and take another breath.
> — Start with three series of 10 movements, gradually building up to three series of 108 movements.

This variation can also be practiced with less intensity. Allow yourself a thin stream of air while inhaling and exhaling during all the aforementioned movements. This makes it possible to do the exercise without kumbhaka.

Nauli Kriyâ

This practice cleans the stomach, intestines and colon through intense internal self-massage. This is done by kneading the stomach through a contraction of the abdominal muscles, in every direction, so that the anterior wall appears to be experiencing a rolling wave of contractions.

Nauli kriyâ, one of the six classic Shatkarmas is, like Agni sara kriyâ, an extension of the practice of Uddîyânabandha. It is easier to accomplish Nauli kriyâ if you have mastered Uddîyânabandha. The practice can be performed in a seated or standing position.

> **Protocol**
> — Place your hands on your thighs and lean slightly forward.
> — Forcefully expel the air from your lungs and, in kumbhaka (holding the breath) pull the abdominal wall back and up, through negative pressure, producing Uddîyânabandha.
> — The first movement begins near the ribs and unfurls toward the pubis, then back, from the pubis to the sternum. Do this a dozen times and then take a few breaths.

> — In the second movement, resume Uddîyânabandha and contract only the abdomen's two large vertical muscles *(Rectus abdominis)*, in the center. This will make them ripple vertically between the sternum and pubis.
> — By pressing a little harder on one thigh, the rectus abdominis muscle on that side will be engaged. Do the same on the other side. This will produce alternating pressure, causing a rotary effect that kneads the stomach from left to right and vice versa. Repeat 10 to 15 times.

This variation can also be practiced with less intensity. Allow yourself a thin stream of air while inhaling and exhaling during all the aforementioned movements. This makes it possible to do the exercise without kumbhaka.

Basti

Basti cleanses the system via colonic irrigation or enema. This practice was originally done standing in a riverbed. Today, enemas are performed with a whole series of flexible tubes and containers of saline water. Again, these are practices that should be supervised by someone with experience.

Asanas of the Abdominopelvic Region

All âsanas that involve the sub-diaphragmatic trunk will loosen the joints of the lumbar vertebrae and pelvis (sacroiliac and pubis). These âsanas will also benefit the superficial and deep paravertebral muscles, psoas muscles and pelvic floor muscles and their related nerves.

Here are some âsanas for the abdominopelvic region: Sîrsâsana, Sarvangâsana, Halâsana, Sasangâsana, Viparîta-karanî, *Uttânâsana (standing forward bend), Pârsvottânâsana (intense side stretch pose), Prasarîta pâdottânâsana (wide-legged standing forward bend), Upavistha konâsana (wide-angle seated forward bend),*

Pascimatânâsana (posterior stretching pose), Karnapîdâsana (ear pressure pose), Kûrmâsana (tortoise), Setu bandhâsana (bridge), Navâsana (boat), Baddha konâsana (cobbler's pose), Ustrâsana (camel), Ekapâda kapotâsana (pigeon), Supta virâsana (reclining hero), Marîcyâsana, Bhujangâsana, Dhanurâsana, Cakrâsana, Trikonâsana and all their variations.

By toning the abdominal corset, Kapâla-bhâti, Breath of Fire and Bhastrikâ prânâyâma have a considerable effect on Samâna vayu (intestinal fire), the Manipura cakra and the Svâdhishthâna cakra.

Mûlabandha and Uddîyânabandha also have similar effects, by toning the pelvic floor and abdominal muscles, especially during âsana practices. A large number of Yoga masters claim that their subtle effects on prânic fluidity and awaking Kundalinî are major triggers.

Freeing the Central Axis

When we look at the physical human structure, it is clear that the vital portion of the physical body is located around its central axis, essentially the trunk. In fact, humans can survive the amputation of part or all of a limb, but not a sectioning of the trunk. The body's central axis, unaltered, is therefore necessary for survival.

To regain or maintain health in these sheaths, we need to ensure that strength, flexibility and power can first be expressed in the trunk. When that is accomplished, the periphery will follow.

Based on this premise, allow me to suggest that, before modifying one's health and physical condition, it would be wise to first attain axial musculoskeletal freedom.

Let us explore how to induce this state by concentrating on freeing the "core" of the physical body. The living physical sheath is free when it has no impediments or tension.

Impediments and Tension

An impediment is undue resistance during the sliding of the tissues. During each movement, the different anatomical structures slide over one another, with the fascias as the friction zones. When there is an impediment, the individual feels tension in the body. This tension mainly occurs in the connective tissues since they form the internal and external structure of all the organs as well as fill the spaces in between. The feeling of tension is triggered by the receptors of the sensory nerves, which are found throughout the fibres of the connective tissue and are stimulated by their stiffness and resistance to movement.

Following is a protocol for the release of the human body's central axis prior to Yoga therapy or even Yoga in general. This initial release could be the starting point for a return to better health or physical shape.

The Art of Releasing Physical Tension

Releasing tension in the tissues and organs is a skill that can only be mastered with clinical experience. However, learning this skill is not very complicated. Manual therapy and bodywork practitioners use several tension releasing techniques. One, in osteopathy, is known as the *Reciprocal Tension Technique*. It exists under other names in the various "manual" medicines.

The beauty of this technique is that it not only releases tension in another person's tissue, but can also be successfully applied to your own.

Description of the technique when applied to another person:

— With one hand, apply gentle traction to the skin and underlying tissue, at the extremity of the area in which tension needs to be released.
— Apply similar traction with the other hand, but in the opposite direction, at the other extremity of the area in which tension needs to be released.
— The traction of one hand should be sufficiently strong to be felt in the other hand and vice versa.
— Hold the tension between the hands for several seconds (rarely more than 30, usually 5 to 10 seconds)
— After this time, you will feel a subtle loosening, as if the tightness were yielding to the gentle traction, as if the tissues were melting. That is the release of tension.

Please note that this release is very subtle and requires very close attention to be perceived. Moreover, practitioners can only master this technique through thousands of repetitions. Not only do you need to know how to apply this reciprocal tension, you need to know the right tissue depth, at which level the tension resides.

Self-Release and Hatha Yoga

You can apply the preceding technique to yourself, using gravity and different Hatha Yoga poses.

Releasing tension with an âsana is relatively easy. Follow the basic rules for even breathing and correct alignment of the structures. Applying these principles to the appropriate poses creates conditions that will encourage this release.

For example, if you want to release posterior cervical tension, you can choose a pose that stretches that area. Several âsanas can apply; the shoulderstand, plough pose, bridge, etc. Once in the desired pose, right at the threshold of tension, simply apply a cocontraction of the area to be treated. Take a few deep breaths, visualizing tension being release with each exhalation. When you release the cocontraction, you will notice that the area is more relaxed. This is because reciprocal auto-balancing of the tension has just occurred.

When tension is released in the physical body, it is usually in the connective tissue. The body's fluids in the area are then free to flow at their maximum efficiency. Prâna (the vital fluid) is also able to circulate fluidly, free from obstructions.

The series of âsanas taught in most Hatha Yoga schools amply covers all areas of the body. This is why diligently applying one of these protocols invariably results in a relaxed physical structure that is filled with energy.

Teaching and Autonomy

Yoga therapy's aim is to help achieve inner harmony and health.

To reach this goal, Yoga therapists must encourage repetition of Yoga's therapeutic sequences. Their clients need to apply the protocol at home and evaluate their progress on the path to well being.

That is why this book was primarily written for Yoga practitioners and health professionals. Clinical experience and teaching experience are major tools in supporting self-healing.

My own clinical experience has taught me that for a therapy to be effective, the patient needs to grasp it in the first session or sessions. Self-empowerment prevents patients from becoming dependent on the therapy or therapist.

Post-Graduate Course for Yoga Therapists

I hereby propose a post-graduate course to be given in various ashrams and Yoga therapy schools. In a few short days, Yoga therapists can learn simple, safe and powerful self-help methods for releasing tension from the central axis of the physical sheath as well as some Yoga-based hygiene practices. They will apply this knowledge first to themselves as they learn and then, if applicable, use it to teach and help their clients.

Epilogue

I am a yogi, a Kriyâ yogi, and I have had a recurring dream for some time:

I am in a little verdant valley in the Himalayas, sitting in siddhâsana on the flowery ground, at the feet of Babaji Nagaraj, my master. I am seated to his right. He is sitting cross-legged on a small, slightly raised rocky platform.

The sky is an intense azure blue, carved by the majestic snowy peaks that surround us. The breeze is pure and cool, making us appreciate the morning sun shining down on us.

Next to me and in front of us, a group of disciples are gathered, also seated on the ground. They form a semi-circle in front of the Master. We are talking to one another, but not out loud. We are actually meditating. When I ask a question, the answer comes instantly, appearing first in my heart, then my mind.

I have had this same dream, on and off, for over 30 years.

Babaji speaks:

Life is expressed by movement.

Where there is movement, there is resistance.

In this dual Nature, there can be no movement without resistance.

An object in movement experience friction with the air, the ground, obstacles. This resistance constantly alters its movement.

Therefore, in Nature, a form of resistance accompanies all movement.

If there is no resistance, there is no movement. That is stagnation.

If there is no movement, there is no life.

To move forward, you must suffer resistance.

If there is no resistance, you do not move forward.

Life therefore cannot express itself if not accompanied by resistance.

It is one of the explanations of Buddha's maxim: "Life is dukkha"; meaning that life is imperfection, dissatisfaction, pain, resistance, suffering.

Sages and rishis have always sought practical solutions to escape this inevitable conditioning. These include Tantra Yoga's yogis and yoginîs.

These adepts propose to transcend the being's classical limitations through the diligent practice of Yoga's psychophysical exercises.

Mastering these exercises eliminates resistance in the physical and mental structures. The result is that the person experiences an extreme state of prânic fluidity.

In modern physics, there are states of matter in which atoms experience no friction or resistance. This is called superfluidity or supraconductivity.

In these conditions, resistance to the flow of particles or charges is practically nil.

Yoga refines the physical and mental structures to the point of perfection equivalent to the supraconductivity of certain materials.

Prânâyâmas, combined with dhyâna (concentration and meditation), pratyâhâra (withdrawal of the senses) and âsanas, are the tools of choice for this task.

In so doing, yogis or yoginîs achieve an ultimate state in which their body is invincible and their mind operates in the supramental spheres characterized by optimal prânic supraconductivity.

In this state, everything is within reach, nothing resists. The experience is one of continual, serene joy.

References

Yoga and Tantra

BOCCIO, Frank Jude, *Mindfulness Yoga*, Boston, Wisdom Publications, 2004, 341 p.

DESIKACHAR, T. K. V., *The Heart of Yoga*, Rochester VT., Inner Traditions, 1999, 244 p.

ELIADE, Mircea, *Patanjali et le Yoga*, Paris, Seuil, 1973, 185 p.

FEUERSTEIN, Georg, *The Yoga Tradition*, Prescott AZ., Hohm Press, 2008, 510 p.

_____*Tantra, The Path of Ecstasy*, Boston, 1998, 316 p.

_____*The Yoga Sûtras of Patanjali*, Rochester Vt., Inner Trad., 1990, 316 p.

GOVINDAN, M. S., *Kriya Yoga Sûtras of Patanjali*, Bolton Qc., KY Pub., 2000, 320 p.

IYENGAR, B.K.S., *Yoga Dipika, Lumière sur le Yoga*, Paris, Buchet-Chastel, 1997, 600 p.

_____*Light On Prânâyâma*, New York, Crossroad Publishing, 2011, 296 p.

JOIS, K. Pathabi, *Yoga Mala*, New York, Patanjali Yoga Shala, 2000, 140 p.

RAMAIAH, S. A. A., *Private Teaching Notes*, Montreal, 1981.

_____*The Voice of Babaji*, Bolton QC., KY Pub., 2006, 524 p.

ROSEN, Richard, *Pranayama, Beyond the Fundamentals*, Boston, Shambhala, 2006, 215 p.

SVÂTMÂRÂMA, *Hathapradîpikâ*, Lonavla, Lonavla Yoga Institute, 2001, 388 p.

THOMPSON, C. G., *Méditation et Ascèse Mantrique*, St-Zénon, Louise Courteau Éditrice, 2007, 373 p.

TIRUMÛLAR, *Tirumandiram, 10 volumes*, Qc., Kriya Yoga Pub., 2010, 3667 p.

VISHNU-DEVANANDA, Swami, *The Complete Illustrated Book of Yoga*, New York, Rivers Press, 1988, 363 p.

Yoga Therapy, Anatomy and Yoga, Alternative Medicine

BATMANGHELIDJ, F, *Your Body's Many Cries for Water*, Falls Chuch VA., Global Health Solutions, 2006, 190 p.

COUDRON, Lionel, *La Yoga-Thérapie*, Paris, Éditions Odile Jacob, 2010, 304 p.

COULTER, D., *Anatomy of Hatha Yoga*, Honesdale PA., Body Breath, 2001, 623 p.

KAMINOFF, Leslie, *Yoga Anatomy*, Champaign IL., Human Kinetics, 2007, 221 p.

KRAFTSOW, Gary, *Yoga for Wellness*, New York, Penguin-Arkana Books, 1999, 334 p.

LIPTON, Bruce, *The Biology of Belief*, Santa Rosa CA., Mountain of Love, 2005, 224 p.

LONG, Ray, *The Key Muscles of Yoga*, Plattsburgh, Bandha Yoga, 2005, 239 p.

_____*The Key Poses of Yoga*, Plattsburgh, Bandha Yoga, 2008, 213 p.

_____*Vinyasa Flow and Standing Poses*, Plattsburgh, Bandha Yoga, 2010, 224 p.

_____*Hip Openers and Forward Bends*, Plattsburgh, Bandha Yoga, 2011, 224 p.

_____*Backbends and Twists*, Plattsburgh, Bandha Yoga, 2011, 224 p.

_____*Arm Balances and Inversions*, Plattsburgh, Bandha Yoga, 2010, 224 p.

OSHMAN, James, *Energy Medicine in Therapeutics and Human Performance*, New York, Butterworth-Heinemann, 2006, 359 p.

ROBIN, Mel, *A Physiological Handbook for Teachers of Yogâsana*, Tucson AZ., Fenestra Books, 2002, 629 p.

SCHLEIP, R., and al., *Fascia : The Tensional Network of the Human Body*, London, Churchill Livingstone, 2012, 566 p.

STILES, Mukunda, *Structural Yoga Therapy*, Boston, Weiser Books, 2000, 344 p.

VARELA, F., and al., *L'inscription corporelle de l'esprit*, Paris, Seuil, 1993, 382 p.

_____*On Becoming Aware*, Philadelphia, J B Publishing, 2003, 245 p.

WÖRLE, Luise, *Yoga as Therapeutic Exercise,* London, Ch. Livingstone, 2010, 256 p.

YOGENDRA, Shri, *Yoga Personal Hygiene*, Bombay, Yoga Institute, 1931, 300 p.

Anatomy and Physiology

GRAY, Henry, *Gray's Anatomy,* London, Churchill Livingstone, 2008, 1576 p.

GUYTON, Arthur C., *Textbook of Medical Physiology,* Philadelphia, Saunders, 2011, 1120 p.

KAHLE and al., *Atlas de Poche d'Anatomie, tomes 1, 2 et 3*, Paris, Flammarion, 2007.

KAPANDJI, A. I., *Physiologie articulaire, tomes 1, 2 et 3*, Paris, Maloine, 2007.

LONGO and al., *Harrison's Principles of Internal Medicine*, New York, McGraw Hill, 2011, 4012 p.

NETTER, Frank, *Atlas d'Anatomie Humaine*, Paris, Elsevier-Masson, 2011, 586 p.

ROBBINS and al., *Pathologic Basis of Disease*, Philadelphia, Saunders, 2009, 1464 p.

Some Websites

All these sites contain useful resources on Yoga and human potential enhancement.

www.aypsite.org

www.babajiskriyayoga.net

www.eomega.org

www.esalen.org

www.himalayaninstitute.org

www.kripalu.org

www.mokshayoga.ca

www.mountmadonna.org

www.rainbowbody.org

www.sivananda.org

www.swamij.com

www.ted.com

www.yogamasters.net

www.ingramcontent.com/pod-product-compliance
Lightning Source LLC
Chambersburg PA
CBHW061152220326
41599CB00025B/4450